Advances and Applications in Deep Eutectic Solvents Technology

Advances and Applications in Deep Eutectic Solvents Technology

Editor

Matteo Tiecco

MDPI • Basel • Beijing • Wuhan • Barcelona • Belgrade • Manchester • Tokyo • Cluj • Tianjin

Editor
Matteo Tiecco
Università degli Studi di Perugia
Italy

Editorial Office
MDPI
St. Alban-Anlage 66
4052 Basel, Switzerland

This is a reprint of articles from the Special Issue published online in the open access journal *Materials* (ISSN 1996-1944) (available at: http://www.mdpi.com).

For citation purposes, cite each article independently as indicated on the article page online and as indicated below:

LastName, A.A.; LastName, B.B.; LastName, C.C. Article Title. *Journal Name* **Year**, *Volume Number*, Page Range.

ISBN 978-3-0365-3563-0 (Hbk)
ISBN 978-3-0365-3564-7 (PDF)

© 2022 by the authors. Articles in this book are Open Access and distributed under the Creative Commons Attribution (CC BY) license, which allows users to download, copy and build upon published articles, as long as the author and publisher are properly credited, which ensures maximum dissemination and a wider impact of our publications.

The book as a whole is distributed by MDPI under the terms and conditions of the Creative Commons license CC BY-NC-ND.

Contents

About the Editor vii

Preface to "Advances and Applications in Deep Eutectic Solvents Technology" ix

Edyta Słupek, Patrycja Makoś-Chełstowska and Jacek Gębicki
Removal of Siloxanes from Model Biogas by Means of Deep Eutectic Solvents in Absorption Process
Reprinted from: *Materials* **2021**, *14*, 241, doi:10.3390/ma14020241 1

Alberto Mannu, Francesca Cardano, Salvatore Baldino and Andrea Fin
Behavior of Ternary Mixtures of Hydrogen Bond Acceptors and Donors in Terms of Band Gap Energies
Reprinted from: *Materials* **2021**, *14*, 3418, doi:10.3390/ma14123418 21

Matteo Ciardi, Federica Ianni, Roccaldo Sardella, Stefano Di Bona, Lina Cossignani, Raimondo Germani, Matteo Tiecco and Catia Clementi
Effective and Selective Extraction of Quercetin from Onion (*Allium cepa* L.) Skin Waste Using Water Dilutions of Acid-Based Deep Eutectic Solvents
Reprinted from: *Materials* **2021**, *14*, 6465, doi:10.3390/ma14216465 29

Alberto Mannu, Marco Blangetti, Salvatore Baldino and Cristina Prandi
Promising Technological and Industrial Applications of Deep Eutectic Systems
Reprinted from: *Materials* **2021**, *14*, 2494, doi:10.3390/ma14102494 47

Kranthi Kumar Maniam and Shiladitya Paul
Ionic Liquids and Deep Eutectic Solvents for CO_2 Conversion Technologies—A Review
Reprinted from: *Materials* **2021**, *14*, 4519, doi:10.3390/ma14164519 73

About the Editor

Matteo Tiecco, PhD earned his degree with full marks in Chemistry from University of Perugia, Italy, in the Dept. of Chemistry. He received his PhD from the same institution (Prof. Savelli). He spent his post-doctoral research activity with Prof. Cardinali, Prof. Cruciani, Prof. Alonso, Prof. Di Profio and Prof. Germani in Perugia (IT), in Chieti (IT), in Alicante (ESP). He earned qualification as Associate Professor in Organic Chemistry in 2019. His research activities are related to surfactants and Deep Eutectic Solvents, discovering the mechanisms of enantioselectivity made by the green solvents themselves by the use of chiral ones. Dr Tiecco earned two EurJOC cover pages; he organized two international conferences; he is a reviewer for Angewandte Chemie, ChemComm, Journal of Molecular Liquids and other journals; he participated in over 36 international and national conferences and he published two books. He was involved in two funded PRIN projects, he won four projects at the synchrotrons of Grenoble (FRA) and Oxford (ENG); he is involved in international collaborations in Australia, Spain, Portugal, the USA, Algeria and Italy in many different Departments and towns.

Preface to "Advances and Applications in Deep Eutectic Solvents Technology"

The development of new technologies and chemical applications nowadays cannot ignore their impact on the environment, due to the widespread diffusion of environmental pollution and the relative consequences also on human health. In this context, the development of novel non-harmful and green solvents represents an excellent contribution to the cause because commonly used solvents are generally toxic, highly volatile, and hardly biodegradable.

Deep eutectic solvents (DESs) represent a new class of green organic solvents that possess the characteristics that can adequately address this problem. Synthesised via combination through weak interactions of two or more solid substances and without the use of any other solvent, DESs possess many green properties: bio-availability of their components; easy biodegradation; low or absent toxicity as well as low or absent vapor pressure. Moreover, they are able to perform catalytic tasks as the properties of the components reflect on the properties of the mixtures obtained.

The focus of this Special Issue is the use of DESs as alternatives to commonly used organic solvents in different areas such as chemical transformations; extraction/preconcentration procedures; electrochemistry; fundamental structural research, and various other topics in which these solvents are finding fruitful applications for the transition from academia to industrial use.

Matteo Tiecco
Editor

Article

Removal of Siloxanes from Model Biogas by Means of Deep Eutectic Solvents in Absorption Process

Edyta Słupek, Patrycja Makoś-Chełstowska * and Jacek Gębicki

Department of Process Engineering and Chemical Technology, Faculty of Chemistry, Gdansk University of Technology, G. Narutowicza St. 11/12, 80-233 Gdańsk, Poland; edyta.slupek@pg.edu.pl (E.S.); jacek.gebicki@pg.edu.pl (J.G.)
* Correspondence: patrycja.makos@pg.edu.pl; Tel.: +48-508-997-100

Abstract: The paper presents the screening of 20 deep eutectic solvents (DESs) composed of tetrapropylammonium bromide (TPABr) and glycols in various molar ratios, and 6 conventional solvents as absorbents for removal of siloxanes from model biogas stream. The screening was achieved using the conductor-like screening model for real solvents (COSMO-RS) based on the comparison of siloxane solubility in DESs. For the DES which was characterized by the highest solubility of siloxanes, studies of physicochemical properties, i.e., viscosity, density, and melting point, were performed. DES composed of tetrapropylammonium bromide (TPABr) and tetraethylene glycol (TEG) in a 1:3 molar ratio was used as an absorbent in experimental studies in which several parameters were optimized, i.e., the temperature, absorbent volume, and model biogas flow rate. The mechanism of siloxanes removal was evaluated by means of an experimental FT-IR analysis as well as by theoretical studies based on σ-profile and σ-potential. On the basis of the obtained results, it can be concluded that TPABr:TEG (1:3) is a very effective absorption solvent for the removal of siloxanes from model biogas, and the main driving force of the absorption process is the formation of the hydrogen bonds between DES and siloxanes.

Keywords: absorption; biogas; deep eutectic solvents; siloxanes

Citation: Słupek, E.; Makoś-Chełstowska, P.; Gębicki, J. Removal of Siloxanes from Model Biogas by Means of Deep Eutectic Solvents in Absorption Process. *Materials* **2021**, *14*, 241. https://doi.org/10.3390/ma14020241

Received: 11 December 2020
Accepted: 31 December 2020
Published: 6 January 2021

Publisher's Note: MDPI stays neutral with regard to jurisdictional claims in published maps and institutional affiliations.

Copyright: © 2021 by the authors. Licensee MDPI, Basel, Switzerland. This article is an open access article distributed under the terms and conditions of the Creative Commons Attribution (CC BY) license (https://creativecommons.org/licenses/by/4.0/).

1. Introduction

The production of energy from renewable sources is not only a choice resulting from the policy of environmental protection or care of the environment but is also an obligation imposed by the European Union in the form of numerous ordinances and international agreements [1]. Therefore, more and more EU countries are focusing their attention on managing waste materials from various industries for the production of biogas [2–5]. This approach is consistent with the theory of sustainable development. However, the obtained biogas is usually a multicomponent mixture containing both inorganic and organic substances, i.e., methane (30–60% v/v), carbon dioxide (15–30% v/v), water, ammonia, hydrogen sulfide, organosulfur compounds, siloxanes, and other linear and aromatic volatile organic compounds (VOCs) [6,7].

The chemical composition of the waste biogas changes depending on the type of raw materials used in the dark fermentation process. The presence of gaseous substances other than methane causes many technological and environmental problems. Particularly dangerous pollutants include siloxane compounds, which can appear in the biogas from municipal landfills or wastewater treatment plants [8,9]. During the combustion of such types of biogas, silicone may be released and combined with oxygen. This can lead to the formation of silica deposits. The silica deposits can cause abrasion of engine parts or the formation of layers that inhibit thermal conductivity or lubrication and clogged transmission lines [10]. Therefore, in order to eliminate the failure of engines converting biogas into energy and to meet the quality requirements for fuels, raw biogas must undergo several treatment processes. The oldest and most widely used process for the treatment of gaseous streams is the application of water or amine scrubbers [11,12]. However, most siloxanes are hydrophobic, and only some of them, i.e., trimethylsilanol, can be absorbed with water because of their high solubility therein [13,14]. Amine scrubbers do not show satisfactory efficiency of siloxane removal either. Among the effective absorbents, there are mineral oils, mixtures of glycols, or inorganic acids [15–18]. Although the above-mentioned absorption methods allow for the recovery of solvents, these methods have a significant disadvantage, which is their energy consumption resulting from the large amount of energy needed to regenerate the absorbent. Therefore, in recent years, more and more scientific research has been devoted to the search for new "green solvents" that will have higher purification efficiency of biogas streams with a simultaneous lower energy demand during regeneration [19].

In the last few years, ionic liquids (ILs) have attracted a lot of attention because they belong to the class of new solvents with a high affinity for CO_2 and a wide range of VOCs [20,21] In addition, ILs have a lower degradation rate, a lower energy requirement for solvent regeneration, and lower corrosive characteristics compared to conventional amine-based solvents [22]. The main disadvantages of ILs are their high viscosity, very high prices, and toxic character. Therefore, deep eutectic solvents (DESs) are a good alternative to ILs because they are much cheaper, less toxic, and more biodegradable [23]. These advantageous properties have made DESs widely used in various separation processes such as extraction [24–27], absorption [28–33], or adsorption [34]. So far, DES has not been used for the experimental removal of siloxanes from biogas. Only theoretical studies can be found in the literature [35].

The study presents screening of twenty-five deep eutectic solvents composed of tetrapropylammonium bromide (TPABr) as hydrogen bond acceptor (HBA) and glycols as hydrogen bond donors (HBDs) in various molar ratio as absorbents for removal of siloxanes from model biogas stream. For this proposal, the conductor-like screening model for real solvents (COSMO-RS) was used. The selection of DESs with the highest siloxane capacity potential was made on the basis of the calculated solubility. For DES (TPABr:TEG 1:3), which was characterized by the highest solubility of siloxanes, the study of its physicochemical properties, i.e., viscosity, density, and the melting point, was performed. Further on, optimization studies of the main parameters influencing the absorption processes were carried out. The mechanism of siloxane removal was evaluated by means of an experimental FT-IR analysis as well as theoretical studies based on σ-profile and σ-potential. To the best of our knowledge, this is the first study dedicated to the application of DES for experimental removal of siloxanes from the gas steams.

2. Materials and Methods

2.1. Materials

The following pure substances were used in this study: tetrapropylammonium bromide (TPABr) (purity $\geq$ 99.0%), tetraethylene glycol (TEG) (purity 99%), hexamethyldisiloxane (L2) (purity 98.5%), octamethyltrisiloxane (L3) (purity 98.5%), and octametylocyclotetrasiloxane (D4) (purity 98%) were purchased from Sigma Aldrich (St. Louis, MO, USA).

For the preparation of model biogas, compressed gases such as nitrogen (purity N 5.5) and methane (purity N 5.0) (Linde Gas, Łódź, Poland) were used. Additionally, for the GC analysis, compressed gases such as nitrogen (purity N 5.5), air (purity N 5.0) generated by a DK50 compressor with a membrane dryer (Ekkom, Cracow, Poland), and hydrogen (purity N 5.5) generated by a 9400 Hydrogen Generator (Packard, Detroit, MI, USA) were used.

2.2. Apparatus

The purification process was controlled by gas chromatography (Autosystem XL) (PerkinElmer, Waltham, MA, USA) coupled with a flame ionization detector (FID) (PerkinElmer, Waltham, MA, USA) and an HP-5 (30 m × 0.25 mm × 0.25 µm) capillary column (Agilent

Technologies, Santa Clara, CA, USA). In the investigations, the TurboChrom 6.1 software (PerkinElmer, Waltham, MA, USA), was used.

The following apparatus was used to evaluate the physicochemical properties: Bruker Tensor 27 spectrometer (Bruker, Billerica, MA, USA) with an ATR accessory and OPUS software (Bruker); BROOKFIELD LVDV-II + viscometer (Labo-Plus, Warsaw, Poland); DMA 4500 M (Anton Paar, Graz, Austria).

2.3. Procedures

2.3.1. COSMO-RS Studies

The geometry optimization of TPABr:TEG (1:3) was performed by means of the continuum solvation COSMO model at the BVP86/TZVP level of theory. The level of theory was used based on previous studies [35,36]. Multiple starting geometries of TPABr:TEG (1:3) were created and optimized in the gas phase to identify stable conformers. In the next step, the vibrational analysis was conducted to find the DES conformer correspond to the true energy minimum. Full geometry optimization was performed only for the most energetically favorable conformer.

In the studies, the COSMO-RS model was used for the screening of DESs using ADF COSMO-RS software (SCM, Netherlands). The relative solubility of siloxanes (x_j) in DESs were calculated using Equation (1):

$$log_{10}(x_j) = log_{10}\left[\frac{\exp\left(\mu_j^{pure} - \mu_j^{solvent} - \Delta G_{j,fusion}\right)}{RT}\right] \quad (1)$$

where: μ_j^{pure}—chemical potential of pure siloxanes (J/mol); $\mu_j^{solvent}$—chemical potential of siloxanes at infinite dilution (J/mol); $\Delta G_{j,fusion}$—fusion free energy of siloxanes (J/mol); R—universal gas constant = 8.314 (J/mol·K); T—temperature (K) [37–39].

2.3.2. Preparation of DES

The deep eutectic solvent was successfully synthesized by mixing TPABr and TEG in 1:3 molar ratio, on a magnetic stirrer under 800 rpm, at 80 °C. All components were dried in a vacuum oven before mixing. The mixing process was carried out for half an hour. The resulting liquid DES was left cooling to room temperature (RT).

2.3.3. Preparation of Model Impurities and Biogas

The model impurities were prepared by means of the barbotage process. Pure nitrogen was moved through a vial containing 1 mL of each siloxane. The obtained model impurities were diluted with a nitrogen stream to acquire a suitable concentration of siloxanes (50 mg/dm^3). This is the upper limit of the range of siloxane concentrations which can be identified in biogas [40].

The model biogas stream was prepared in two options. The first with the use of pure nitrogen, and the second with the use of a mixture of nitrogen and methane gases in the volume ratio of 2:1.

2.3.4. Absorption Process

The installation to separate the siloxanes consists of an absorption column, a stripper column, a heat exchanger, and a reboiler. Figure 1 shows the process of the absorption–desorption course of siloxanes using TPABr:TEG (1:3). The model polluted biogas stream containing a certain amount of methane and siloxanes is fed into the absorption column. The absorption process takes place under certain conditions maintained in the column (temperature of the process—Ta, the volume of DES—Va, flow rate of the biogas stream—wa). Pure methane from the top of the absorption column is collected. The next step in the entire process is desorbing the siloxanes from DES. For this purpose, the contaminated DES is directed into the stripper column which works in specific conditions (temperature

of the stripper process—Ts, time of the stripper process—ts). Owing to regeneration, it is
possible to reuse DES, which has a major impact on the economics of the process.

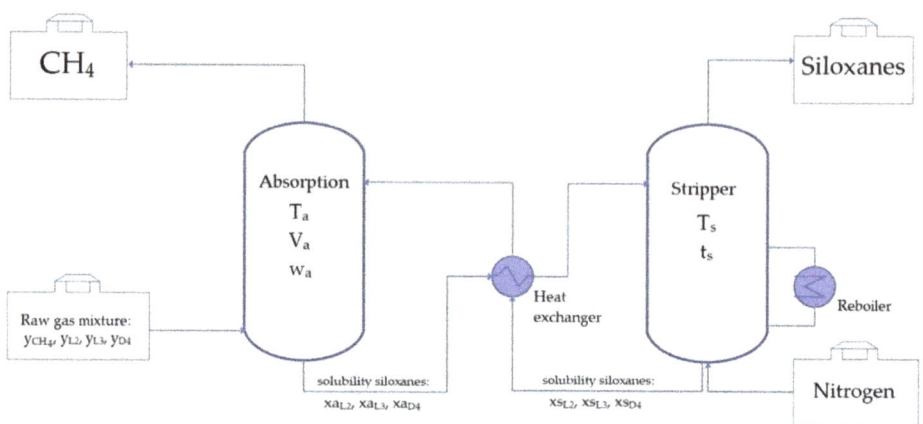

Figure 1. Simplified process flow diagram of siloxanes separation using deep eutectic solvent (DES).

The absorptivity (A) of siloxanes in the TPABr:TEG (1:3) was calculated using Equation (2):

$$A = \frac{C_{in} - C_{out}}{C_{in}} \quad (\text{-}) \tag{2}$$

where c_{in}—initial siloxanes concentration (mg/dm^3), c_{out}—siloxanes concentration after
absorption process (mg/dm^3).

2.3.5. Regeneration of DES

Following the selective absorption process of siloxanes, TPABr:TEG (1:3) was regenerated using nitrogen barbotage at an elevated temperature (80–100 °C). The regeneration
process was carried out conducted in line with previous studies [24]. The regeneration
experiments were conducted at 90 °C with an N$_2$ flow of 50 mL/min. The concentration of
L2, L3, and D4 (before and after regeneration) in TPABr:TEG (1:3) was studied by means of
gas chromatography.

2.3.6. Chromatographic Analysis

The degree and efficiency of the model biogas treatment were determined by gas chromatography coupled with a flame-ionization detector (GC-FID) (PerkinElmer, Waltham,
MA, USA). The temperature of the GC oven was 120 °C, the detector temperature was
300 °C, the injection port temperature was 300 °C, the injection mode was split 5:1, and the
carrier gas was nitrogen (2 mL/min).

2.3.7. Physicochemical Properties of DES
FT-IR Analysis

FT-IR spectra were taken using attenuated total reflectance (ATR) with the following
operating parameters: number of background scans: 256, number of sample scans: 256;
spectral range: 4000–550 cm^{-1}; resolution: 4 cm^{-1}; and slit width: 0.5 cm.

Viscosity and Density Measurements

The viscosity and density of the synthesized TPABr:TEG (1:3) were measured within
a temperature range of 25–60 °C. The uncertainty measurement for the temperature was
0.5 °C. Additionally, the relationship between the viscosity and revolutions per minute
abbreviated (RPM) in the temperature range 25–60 °C was determined.

Melting Point Measurements

The melting point (MP) was determined visually at atmospheric pressure by cooling DES samples to −50 °C, followed by a temperature increase at 0.1 °C/min. The temperature at which the initiation of the phase transformation was observed was adopted as the melting point.

3. Results and Discussion
3.1. COSMO-RS Molecular Simulation
3.1.1. Solubility of Siloxanes in DESs—Preselection of DES

The conductor-like screening model for real solvents (COSMO-RS) was used to calculate the solubility of siloxanes in pure glycols and water and in DESs composed with TPABr and glycols. Based on the previous studies, it can be deduced that COSMO-RS is a useful tool for solvent screening and predicting the impurities' solubility in conventional as well as non-conventional solvents [35,41,42]. In most of the published works, the selection of solvents is made on certain parameters, i.e., Henry's constant and activity coefficient. The results are often inconsistent. However, the most important parameter from the industrial point of view, solubility, is rarely reported [35,43]. Therefore, in this study, we calculated the solubility of individual siloxanes (L2, L3, and D4) in various DESs composed of TPABr as HBA and glycols, i.e., ethylene glycol (EG), glycerol (Gly), triethylene glycol (TriEG), tetraethylene glycol (TEG), and diethylene glycol (DEG), as HBD at various molar ratios (1:3; 1:4; 1:5; 1:6, HBA:HBD). These various molar ratios were selected on the basis of other studies which show that the melting point of most TPABr:glycols in 1:1, 1:2 complexes are higher than 20 °C [44,45]. This fact disqualifies the possibility of such DES as absorbents since one of the necessary conditions for absorbents is liquid at room temperature. The structures of HBA and HBDs are presented in Figure 2.

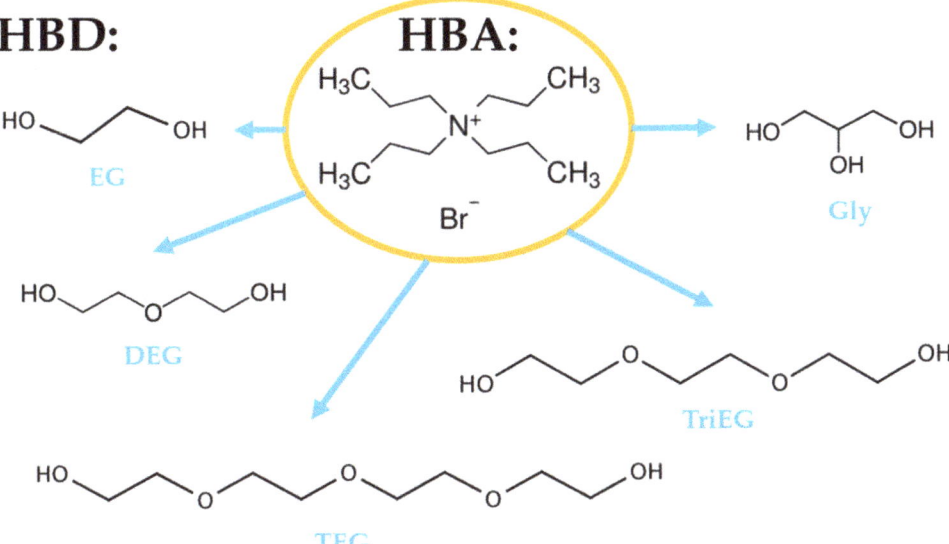

Figure 2. Structures of hydrogen bond acceptor (HBA) and hydrogen bond donors (HBDs).

Additionally, the solubility of siloxanes in pure glycols and water were taken into account. The obtained results are presented in Figures 3–5.

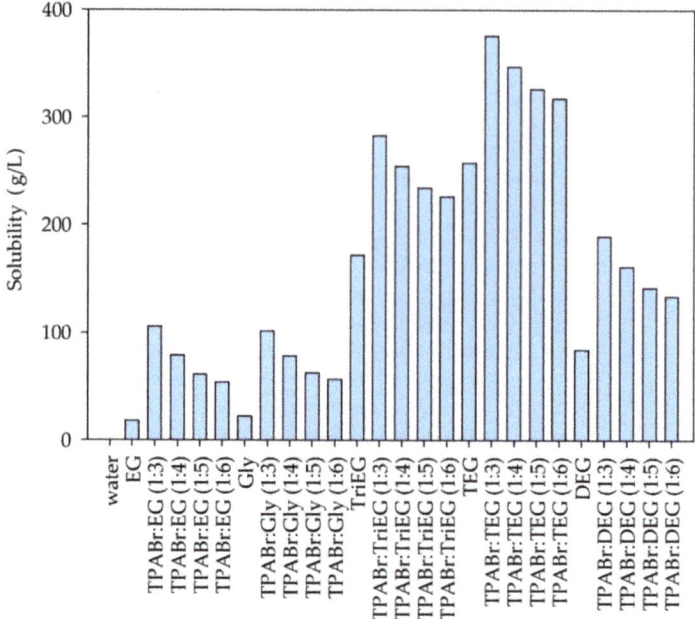

Figure 3. Solubility of hexamethyldisiloxane (L2) in DES and pure solvents.

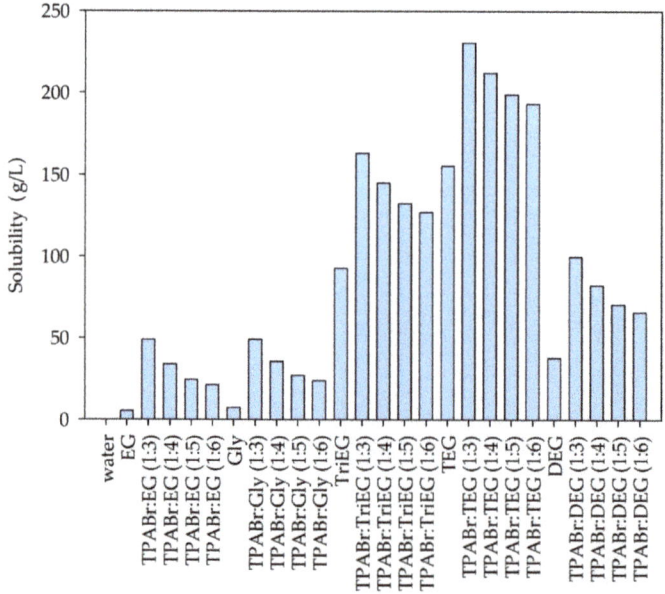

Figure 4. Solubility of octamethyltrisiloxane (L3) in DES and pure solvents.

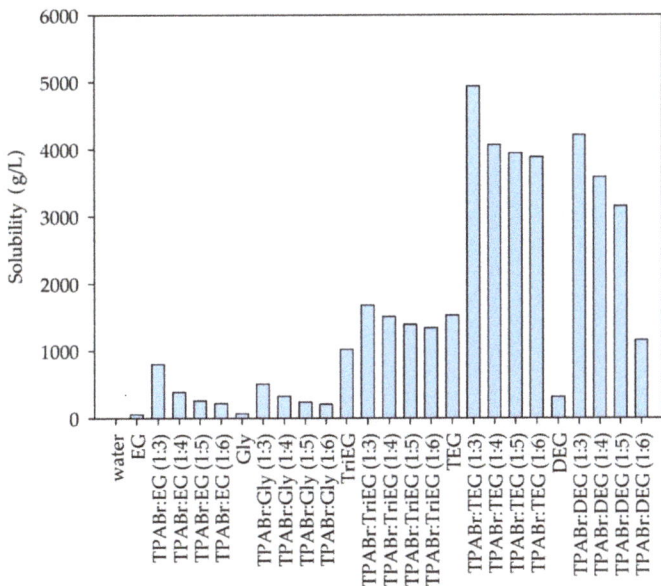

Figure 5. Solubility of octametylocyclotetrasiloxane (D4) in DES and pure solvents.

Among the tested solvents, water is the poorest solvent for siloxanes. The calculated solubility of individual siloxanes in water was 0.0054, 0.00011, and 0.027 g/L for L2, L3, and D4, respectively. This is due to the hydrophobic nature of most siloxanes [13,14]. From the industrial point of view, the ideal absorbents should be cheap and easily accessible. Due to the relatively high price of TPABr in comparison to the price of pure glycols, it would be advantageous to use pure EG, Gly, TriEG, TEG, or DEG to absorb siloxanes from biogas. However, the calculated solubility values are significantly lower for pure glycols compared to DES. The highest solubility can be observed for the DES composed of TPABr and glycols in 1:3 molar ratio. On the other hand, further increasing the amount of glycols in DES structures reduces the solubility of siloxanes. This indicates that both HBA and HBD take an active part in the absorption process by creating hydrogen or electrostatic bonds with siloxanes. COSMO-RS calculations indicate that D4, which represents cyclic siloxanes, shows higher solubility in DESs than linear siloxanes (L2 and L3). Similar results were obtained for ILs in the previous studies [46]. For linear siloxanes, as the length of the molecule decreases, their solubility in DESs increases. These are different results from those obtained for ionic liquids [46]. The highest solubility of both linear and cyclic siloxanes was obtained for DES composed of TPABr and TEG in 1:3 molar ratio. This is probably due to the formation of strong non-bonded interactions between TPABr:TEG (1:3) and siloxanes, i.e., hydrogen bonds between -OH group from TEG molecules, and O—a group from siloxanes. In order to obtain detailed information on the interactions between DES and siloxanes, analyses of σ-profiles and σ-potentials were performed. Due to the best siloxane dissolving ability of TPABr:TEG (1:3), only this DES was further investigated.

3.1.2. σ-Profile

A very important molecule-specific property in the COSMO-RS model is the σ-profile, which is the probability distribution of surface area with charge density (σ). Typically, σ-profile is presented as a histogram which can be divided into three regions i.e., HBA region $\sigma > 0.0084$ e/Å^2; non-polar region -0.0084 e/Å$^2 < \sigma < 0.0084$ e/Å^2; and HBD region $\sigma < -0.0084$ e/Å^2 [47]. The σ-profiles of TPA, Br, TEG, L2, L3, and L4 are shown in Figure 6.

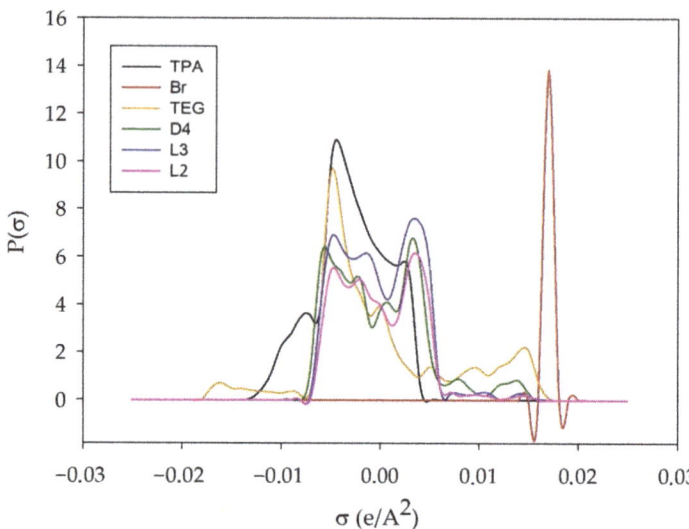

Figure 6. σ-Profile of TPABr:TEG (1:3), L2, L3, and L4.

The results indicate that the σ-profiles of all siloxanes are distributed within the non-polar and hydrogen bond acceptor region. There is no significant difference between σ-profiles of L2, L3, and D4. The only peak can be observed in the more positive region of the histogram for linear siloxanes. This indicates a slightly stronger hydrogen bond acceptor capacity of L2 and L3. Similar results were observed in other studies [35,46]. The distribution of TPA shows the concentration of the charge density mainly around 0, and a small concentration of the charge below −0.0084, which indicates the role of TPA as a hydrogen bond acceptor in hydrogen bond formation. The distribution of the bromide anion is located around 0.018 in the HBA area, which demonstrates a non-polar character and the possibility of H-bonding formation. On the other hand, the distribution of TEG is observed over the entire range of the σ-profile. This indicates that TEG can be both an acceptor and a hydrogen bond donor.

3.1.3. σ-Potential

The σ-potential is typically used to indicate the affinity between mixture components. The higher values of the positive μ (σ) suggest an increase in its repulsive behavior, and higher negative values of the μ (σ) indicate a stronger interaction between the molecules. Similarly to the σ-profile plot, the σ-potential plot is divided at the same three regions. The σ-potential for TPABr:TEG (1:3), L2, L3, and D4 are plotted in Figure 7. The obtained results indicate that all siloxanes almost overlap each other, which means that L2, L3, and D4 have similar molecular interaction properties with other molecules and with themselves. The shape of siloxanes σ-potential is negative in the HBA region and positive in the HBD region. This means that L2, L3, and D4 can be acceptors in H-bonding formation. However, the DES shape is negative in both these regions. This indicates that it is both an acceptor and a hydrogen bond donor. Therefore, the formation of hydrogen bonds is the most likely driving force in the process of removing siloxanes from biogas.

3.2. Structural and Physicochemical Properties of DES

3.2.1. FT-IR Analysis

Spectroscopic characterization is a very important aspect to determine the interaction between HBA (TPABr) and HBD (TEG). For this purpose, the FT-IR analysis was used in the study (Figure 8).

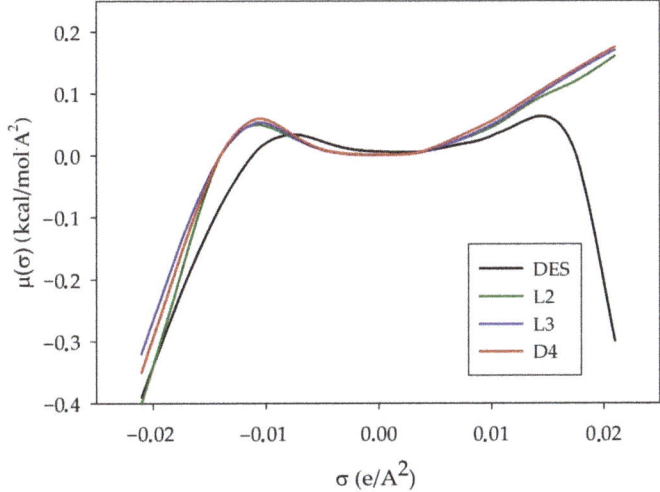

Figure 7. σ-Potential of TPAB:TEG (1:3) L2, L3, and L4.

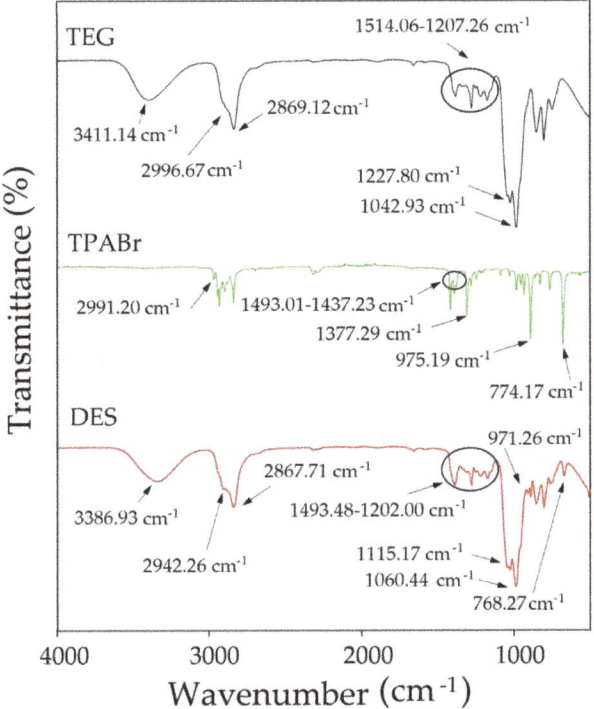

Figure 8. FT-IR spectrum for pure TPABr, TEG, and TPABr:TEG (1:3).

Figure 8 shows the mechanism of TBABr:TEG (1:3) formation. In the TPABr:TEG (1:3) spectrum, the shift of -OH group vibration towards lower values compared with pure TEG HBD (from 3411.14 to 3386.93 cm^{-1}) indicates the formation of O-H O or O-H Cl bonds. In addition, the broadening and shifts of the vibration towards lower values of the aliphatic C–H stretching bonds (from 2996.67 and 2869.12 cm^{-1} to 2942.26 and 2867.71 cm^{-1}) can be observed. The shifts O-H O or O-H Cl, and C-H groups are most likely the consequence of hydrogen bond formation between TPABr and TEG [48,49]. Moreover, shifts of the OH group may result from the presence of C-O-C groups in the TEG. The C-O-C group are considered as the electronegative groups and tend to attract electrons on hydrogen in OH bands. The occurring interactions between TPABr and TEG can be confirmed by the shift of the C-O-C group towards lower values from 1227.80 to 1115.17 cm^{-1} and increasing the intensity of the OH group [50]. Similar vibration towards lower values can be seen in the peaks in the bands responding with H-bending, CH_2 deformation, and N-C-C bending bonds from 1514.06–1207.26 cm^{-1} to 1493.48–1202.00 cm^{-1}, and C-N bond symmetric stretching vibration from 774.17 to 768.27 cm^{-1} as well as redshift phenomena O-H and C-O-H stretching bonds from 1042.73 to 1060.44 cm^{-1}. The shifts confirm the interaction between the atoms in TPABr and TEG [51–53].

3.2.2. Viscosity and Density Measurements

It is well known that DES components and temperature have a dramatic effect on the absorbent density and viscosity. Basic physicochemical parameters of DES strongly influence the ability of the mass transfer capacity, which is of great importance for any changes in the absorption process [54,55]. In order to analyze the flow behavior of synthesized TPABr:TEG (1:3), the viscosity was studied in a function of shear rate ranging 10–50 rpm and temperature range 25–60 °C. The obtained results indicate that the viscosity of TBABr:TEG (1:3) decreases with increasing temperature. The increase in temperature causes the velocity of the particles in the liquid to increase, which reduces the intermolecular forces, resulting in a decrease in the TPABr:TEG (1:3) viscosity (Figure 9A). At room temperature, the viscosity of TPABr:TEG (1:3) is 84.6 mPas; it should be noted that it is much lower compared to the DESs which are presented in the literature. The dynamic viscosity of DES composed of tetrabutylammonium bromide (TBABr) and glycerol (Gly) or ethylene glycol (EG) in a molar ratio of 1:3 were 467.2 and 91.4 mPas, respectively [56]. A decrease in the viscosity value contributes to the increase in the capacity and rate of absorption because it makes the mass transfer easier. Therefore, DESs with lower value viscosities are more desirable for absorption processes.

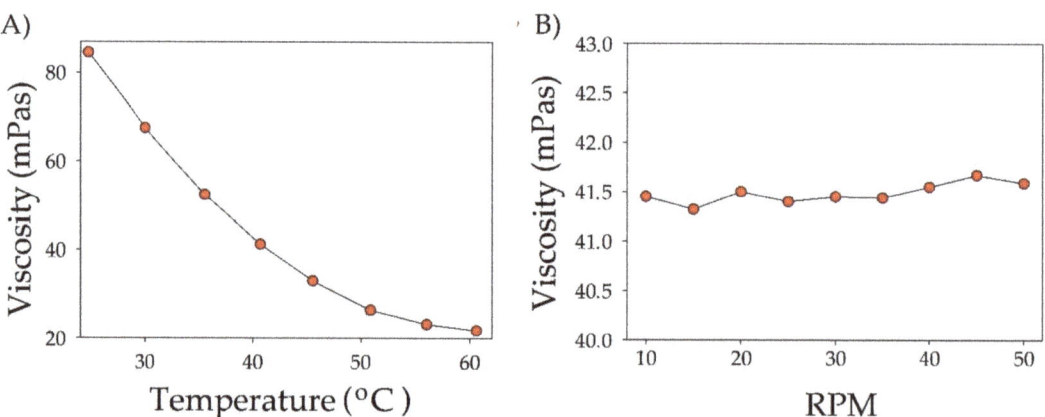

Figure 9. (**A**) Viscosity for TPABr:TEG (1:3) as a function of temperature in the range 25–60 °C. (**B**) Dependence of viscosity on turnover in the range 10–50 RPM in temperature 40 °C.

In Figure 9B, it can be observed that the viscosity of TPABr:TEG (1:3) remains almost constant throughout the range of the applied shear rate ranging. Therefore, it can be concluded that the obtained DES is a Newtonian liquid [57]. The possible shear thinning behavior can be attributed to different strengths of the H-bonding present in TPABr:TEG (1:3) which can start breaking with increasing RPM. However, deeper analysis is required to confirm these assumptions. Similar behavior was also observed for another type of DES [58].

Another tested physicochemical parameter was density. The value of DES density decreases linearly with increasing temperature (Figure 10). At 20 °C, the density TPABr:TEG (1:3) is 1.5520 g/cm^3. However, it can be observed that with increased temperature (60 °C), the density value decreases to 1.5508 g/cm^3. The lower density values can be due to the fact that during heating, HBA and HBD in DES vibrate harder. These vibrates can cause molecular rearrangements between HBA and HBD, which can contribute to creating weaker interactions in the hydrogen bonding [59]. The obtained density of TPABr:TEG (1:3) is higher compared to the DESs which are composed of quaternary ammonium salts (ChCl or TBABr) [56,60].

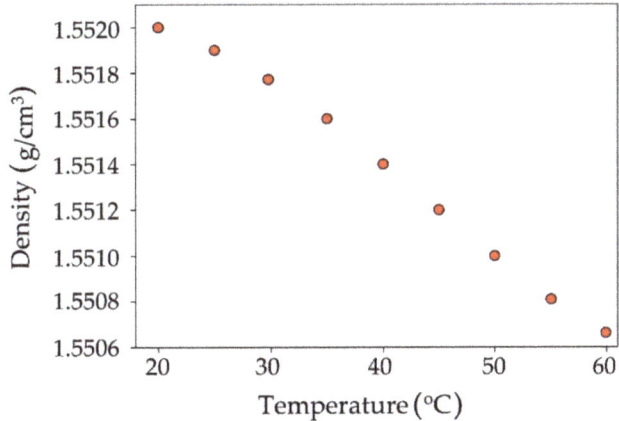

Figure 10. Density for TPABr:TEG (1:3) as a function of temperature in the range 25–60 °C.

3.2.3. Melting Point Measurements

The measured MP of TPABr:TEG (1:3) is −48 °C. As expected, the melting point of TPABr:TEG (1:3) is lower than the MP of pure TEG (−9.4 °C) [61]. The depression in the melting point of the mixture shows the formation of strong intermolecular interactions, i.e., hydrogen bonds between TPABr and TEG [62].

3.3. An Experimental Studies on Absorption of Siloxane Compounds

Optimization of the Absorption Process Conditions

In our research, the processes of absorption using a new DES based on TPABr:TEG (1:3) were carried out for purification of the model biogas stream from L2, L3, and D4 pollutants. The absorption processes were optimized in terms of the volume of TPABr:TEG (1:3), model biogas flow, and temperature.

The first optimized parameter was the volume of the TPABr:TEG (1:3) in the range of 15–50 mL/min (Figure 11). The results show that the volume of DES has a significant impact on the overall siloxane capture process. As the volume of DES increases from 15 to 50 mL/min, the DES saturation time increased from 150 to 320 min (L3—Figure 11B), from 140 to 400 min (L2—Figure 11A), and from 1551 to 5281 min (D4—Figure 11C). The increase in saturation time can be explained by increases in the contact time between the siloxane gas phase and the absorbent [63]. Increasing the volume of DES also contributes to an increase in the amount of active substance (TPABr:TEG (1:3)) and an increase in the number of active sites that are responsible for capturing of the siloxanes from DES.

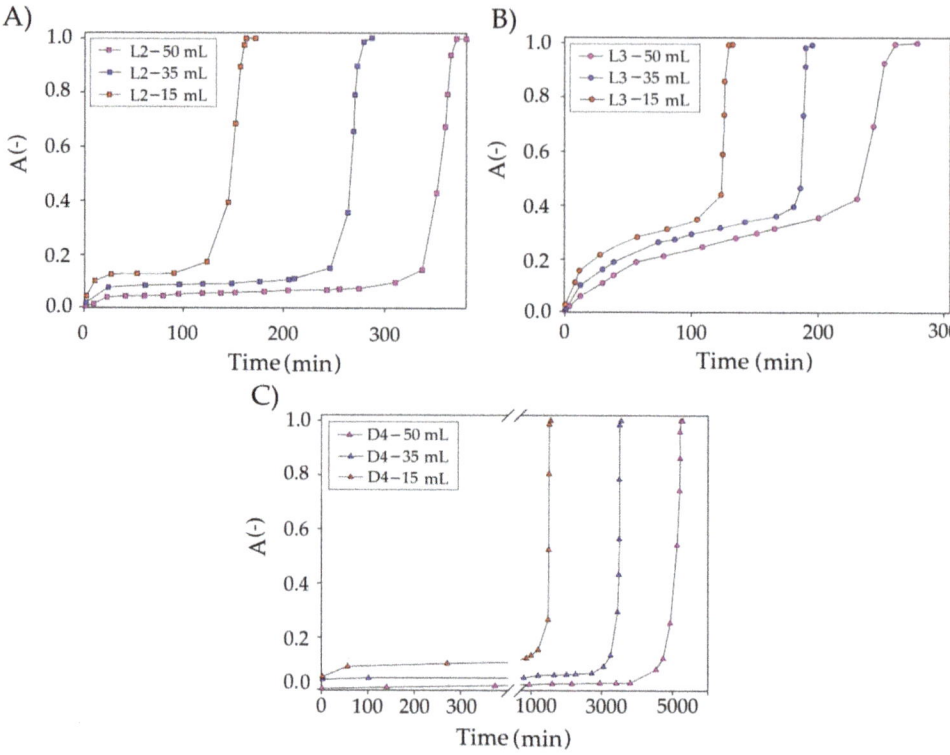

Figure 11. Experimental breakthrough curves of siloxane absorption with TPABr:TEG (1:3) at the different volumes of DES for (**A**) L2; (**B**) L3; and (**C**) D4.

The next studied parameter was model biogas flow rate in the range of 10–50 mL/min (Figure 12). The results indicate that the flow rate has only a minor effect on siloxane uptake compared to DES volume. The conducted research indicates that an increase in the flow rate from 10 to 50 mL/min slightly decreased the effectiveness of siloxane removal from the model biogas stream. Similar results were observed in the previous studies [29,64]. In the industrial technologies used with the use of a water scrubber, a flow of 88 mL/min is used to remove CO_2 or H_2S [65], whereas when using an amine scrubber, flows of 30 mL/min are used [66]. The reduction in the flow rate may result from the different viscosities of the use of the absorbent. Therefore, the assumed optimal value of 50 mL/min seems to be the rational and comparable value.

The temperature in the range of 25–50 °C was selected as the last parameter for optimizing the absorption conditions (Figure 13). An increase in temperature causes decreases in TPABr:TEG (1:3) viscosity. The lower viscosity improves the mass transfer efficiency and, hence, the siloxane removal efficiency should be higher. However, increasing the temperature does not extend the absorption process too much. This is likely due to the fact that the absorption process is normally exothermic [67]. Therefore, a temperature of 25 °C was adopted as optimal.

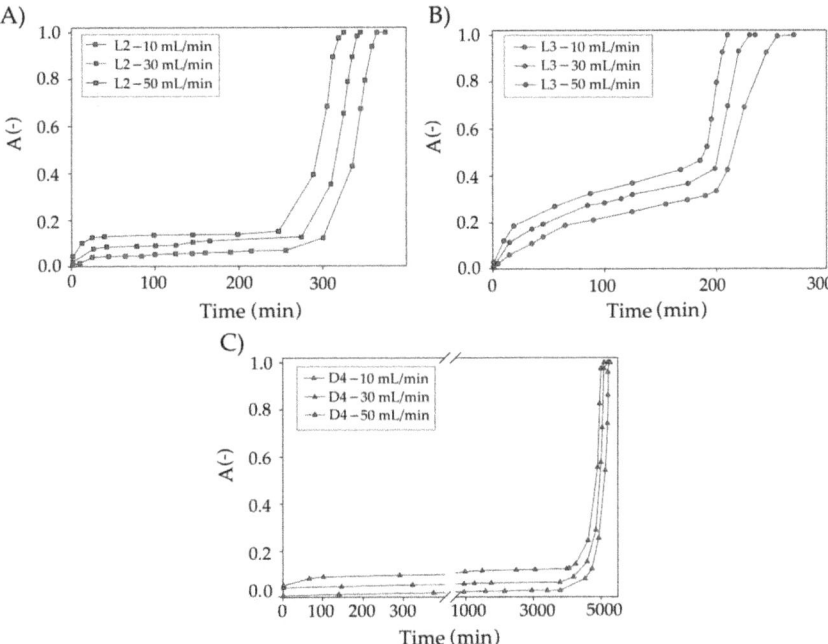

Figure 12. Experimental breakthrough curves of siloxane absorption with TPABr:TEG (1:3) at the different biogas flow rate for (**A**) L2; (**B**) L3; and (**C**) D4.

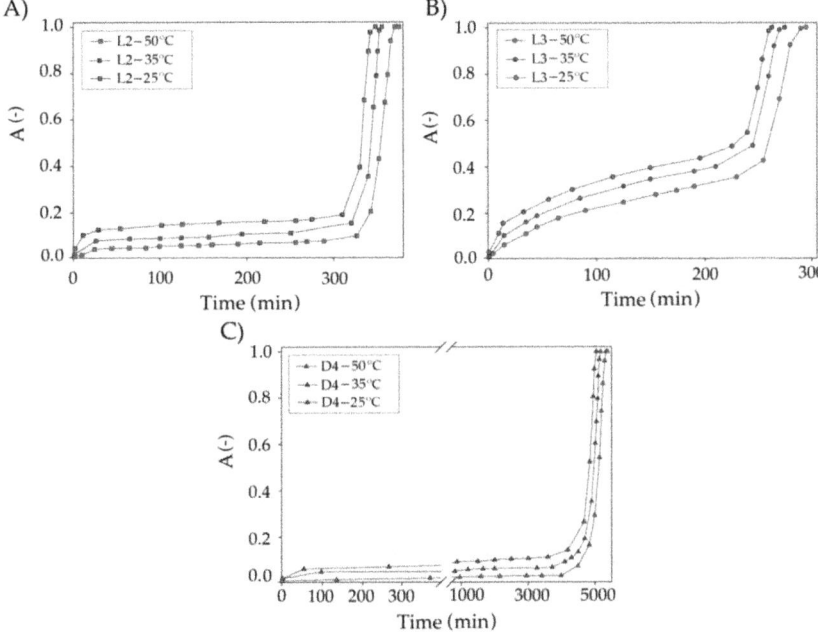

Figure 13. Experimental breakthrough curves of siloxane absorption with TPABr:TEG (1:3) at different temperatures for (**A**) L2; (**B**) L3; and (**C**) D4.

Owing to the conducted research, the optimum conditions for the removal of siloxanes from the model biogas stream were selected as a temperature of 25 °C, a DES volume of 50 mL, and a flow rate of 50 mL/min. The obtained dependence of the absorption efficiency on the duration of the absorption process of individual pollutants is shown in Figure 14A (with the use of pure nitrogen) and Figure 14B (with the use of a mixture of nitrogen and methane gases in the volume ratio 2:1).

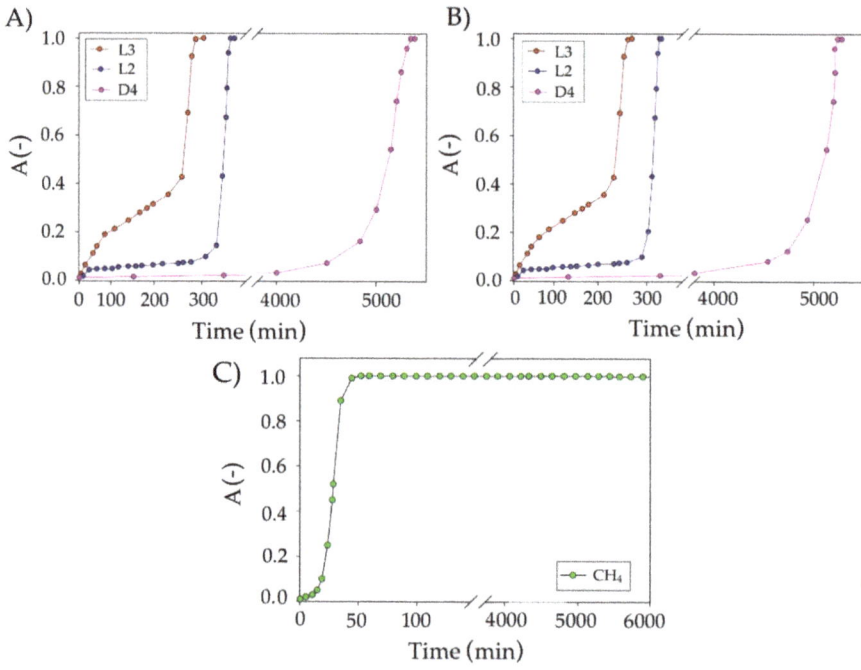

Figure 14. Experimental breakthrough curves of impurities absorption with TPABr:TEG (1:3) using (**A**) pure nitrogen and (**B**) mixture of nitrogen and methane at 2:1 volume ratio under optimum conditions, and (**C**) methane absorption curve using TPABr:TEG (1:3).

For D4 in pure N_2, after 5380 min of absorption process, a sharp increase in the supersaturation value was observed. While for D4 in the mixture of nitrogen:methane at 2:1 volume ratio, the saturation time was 5300 min. The oversaturation times of the other two siloxanes in N_2 were 400 and 300 min, while in N_2:CH_4 (2:1), they were 375 and 280 min, respectively, for L2 and L3. In the literature, there are very few works that focus on the capture of siloxanes from biogas. The results obtained in our studies can only be compared to the absorption in which the absorbent consists of amines, acids, or bases. However, it is well known that the strong bases and acids contribute to the cleavage of Si-O bonds and convert siloxanes to other volatile compounds with lower boiling points [68].

Devia and Subrenat [15] proposed L2 and D4 absorption into motor oil, cutting oil, and water-cutting oil. The studies showed the best results were obtained for motor oil for which the breakthrough curves obtained to allow for efficient removal of siloxanes were for 191.4 min (L2) and for 47.1 min (D4). The obtained results show that the proposed new DES-based absorbents show a much higher absorption capacity towards siloxanes than conventional solvents. In the studies, apart from monitoring the siloxane absorption process, the concentration of methane was also monitored (Figure 14C). The results show that complete saturation of TPABr:TEG (1:3) with methane occurs after 50 min of the process. The loss of methane in the entire process of siloxane absorption was within 5%.

3.4. FT-IR Studies on Absorption of Siloxane Compounds

The experimental study on the mechanism of the absorption process of siloxanes was conducted by FT-IR analysis. The obtained spectra of pure TPABr:TEG (1:3) and pure siloxanes were compared with the spectra of TPABr:TEG (1:3) after the absorption process (Figures 15–17). All of them identified the bands which can be observed in the FT-IR spectrum for pure siloxanes: Si-O-Si antisymmetric stretch bonds in the range 1000–1100 cm^{-1} and Si-C symmetric stretch bonds at 800 cm^{-1} are visible in the spectrum of the TPABr:TEG (1:3) after the absorption process [69]. In the spectrum after the absorption process, new peaks or significant band shifts cannot be observed. Only shifts of the -OH stretching vibration and aliphatic C-H stretching bonds are visible, which confirms the phenomena of physical absorptions. In addition, the shifts of -OH stretching vibration indicate that the hydroxyl group from TPABr:TEG (1:3) may interact with the oxygen atoms from siloxanes by forming hydrogen bonds, which is in accordance with the siloxane absorption [70]. Additionally, a shift of the bands originating from group C-O-C towards higher values from 1112.40 to 1123.83 cm^{-1} (Figure 15), 1123.54 cm^{-1} (Figure 16), and 1118.59 cm^{-1} (Figure 17) are observed. These shifts indicate that siloxane absorption can also occur through the interaction of silicon atom (Si-OH—827.67 cm^{-1} and SiO—752.14 cm^{-1} (Figure 15), Si-O—801.56 cm^{-1} (Figure 16), Si-OH and Si-O in the range 847.96–792.11 cm^{-1} (Figure 17)) with the oxygen atoms with C-O-C in the DES (1:3) [50].

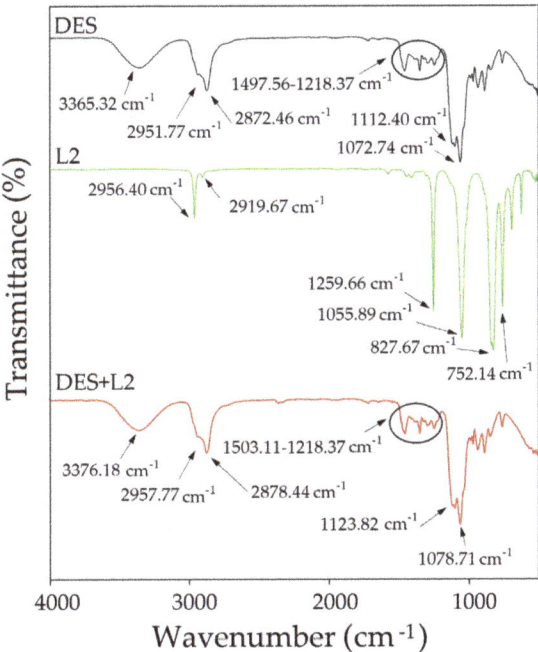

Figure 15. FT-IR spectrum for pure TPABr:TEG (1:3), pure L2 and complex TPABr:TEG + L2 (DES + L2).

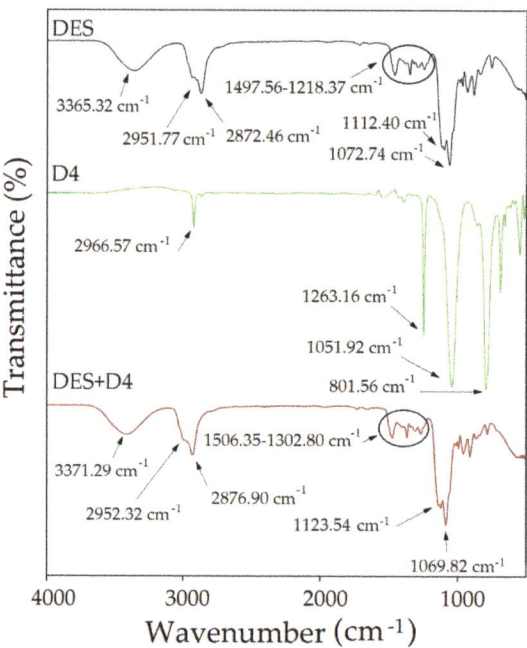

Figure 16. FT-IR spectrum for pure TPABr:TEG (1:3), pure D4 and complex TPABr:TEG + D4 (DES + D4).

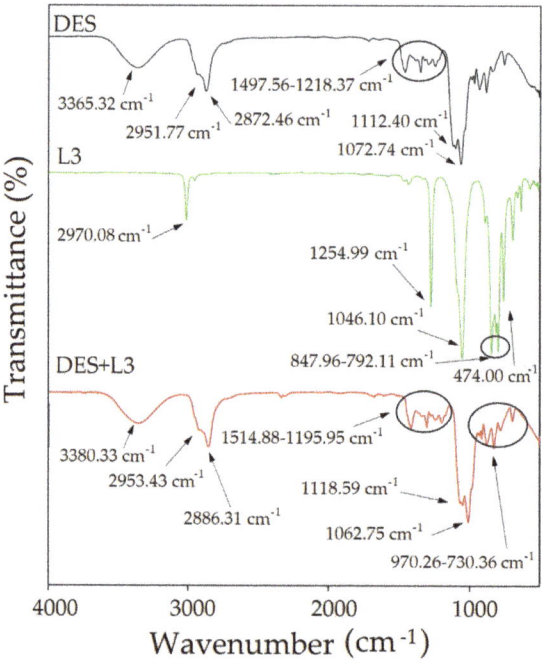

Figure 17. FT-IR spectrum for pure TPABr:TEG (1:3), pure L3 and complex TPABr:TEG + L3 (DES + L3).

3.5. Regeneration of DES

From the industrial point of view, the regeneration processes of absorbents are extremely important because they determine the final costs. The obtained results indicate that siloxanes can be completely removed from TPABr:TEG (1:3) using nitrogen barbotage conducted at a temperature of 90 °C for 3 h. TPABr:TEG (1:3) shows tall and almost unchanging L2, L3, and D4 removal efficiency for up to 10 regeneration cycles (Figure 18A). In addition, the thermal stability of TPABr:TEG (1:3) by means of FT-IR analysis was confirmed. The comparison of fresh and regenerated TPABr:TEG (1:3) spectrum indicates a lack of additional shifts and peaks in the regenerated TPABr:TEG (1:3) (Figure 18B). This confirms stability and effective regeneration of TPABr:TEG (1:3).

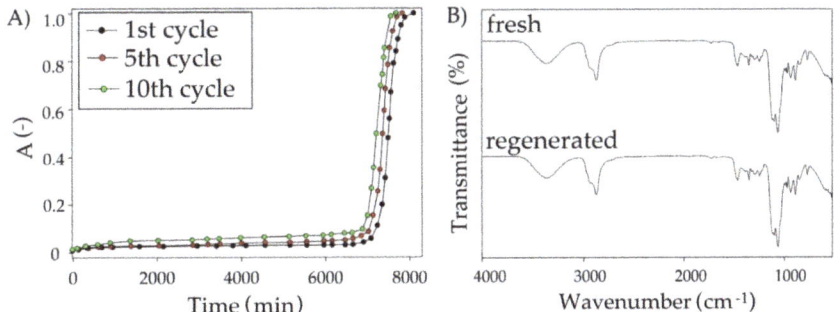

Figure 18. (**A**) Reusability of TPABr:TEG (1:3) on the example of removing L2, (**B**) FT-IR spectra from on fresh and regenerated TPABr:TEG (1:3).

4. Conclusions

In the paper, the solubility of siloxanes (L2, L3, and D4) in deep eutectic solvents (DESs) composed of tetrabutylammonium bromide and glycols as well as conventional solvents was investigated. The siloxane solubility was predicted by means of the COSMO-RS model where the highest solubility of both linear and cyclic siloxanes was present in the DES composed of TPABr and TEG in a 1:3 molar ratio. The chemical structures of TPABr:TEG (1:3) and the interaction structures between TPABr and TEG as well as between DES and siloxanes were reported using FT-IR spectroscopy. Furthermore, in order to confirm the interactions, the analyses of σ-profiles and σ-potentials were used.

The results of physicochemical properties indicate that TPABr:TEG (1:3) is a Newtonian liquid due to the lack of viscosity changes during shear changes, which contribute to only minor changes in siloxane removal efficiency with increasing temperature. In turn, carrying out the absorption process at a temperature of 25 °C is beneficial from an economic point of view. Under optimum conditions (50 mL of TPABr:TEG (1:3), 50 mL/min flow rate, and temperature 25 °C), the L2, L3, and D4 can be removed with high efficiency for 375, 280, and 5300 min, respectively. These are much better absorption efficiencies compared to mineral oils. In addition, TPABr:TEG (1:3) also can be easily regenerated up to 5 cycles without significantly changing the siloxane absorption efficiency. The studies on the absorptive mechanism to remove siloxanes indicate that the reason for the high solubility siloxanes in TPABr:TEG (1:3) is the formation of the strong hydrogen interactions between -OH group from DES molecule, and -O- a group from siloxanes.

The cost of the absorption process mainly depends on the type of absorbents. The estimated capital cost of the absorption process based on TPABr:TEG (1:3) is 126.05 €/L [71,72]. The DES price is higher than conventional absorbents prices which are 9.17 €/L for motor oil 5W40 from Elf) [73]; 33.42 €/L for cutting oil Hochleistungs-Schneidöl Alpha 93 from Jokish® GmbH [74]. However, it should be remembered that DES can be used for up to 5 cycles without changing the high efficiency of removing siloxanes from biogas. The initial price of 126.05 €/L can drop to 25.21 €/L. Therefore, DESs can be used as alternative absorbents.

Author Contributions: Conceptualization, E.S. and P.M.-C.; methodology, E.S. and P.M.-C., investigation, P.M.-C., E.S., and J.G.; data curation, P.M.-C. and E.S.; writing—original draft preparation, P.M.-C. and E.S.; writing—review and editing, J.G.; visualization, E.S. and P.M.-C., supervision, J.G., funding acquisition, P.M.-C. All authors have read and agreed to the published version of the manuscript.

Funding: This research was funded by ARGENTUM TRIGGERING RESEARCH GRANDS program as part of the Excellence Initiative—Research University program within the grant project (No. DEC-34/2020/IDUB/I.3.3).

Institutional Review Board Statement: Not applicable.

Informed Consent Statement: Not applicable.

Data Availability Statement: The data presented in this study are available on request from the corresponding author.

Conflicts of Interest: The authors declare no conflict of interest.

References

1. Perea-Moreno, M.-A.; Salmeron-Manzano, E.; Perea-Moreno, A.-J. Biomass as Renewable Energy: Worldwide Research Trends. *Sustainability* **2019**, *11*, 863. [CrossRef]
2. Meyer, A.K.; Ehimen, E.A.; Holm-Nielsen, J.B. Future European biogas: Animal manure, straw and grass potentials for a sustainable European biogas production. *Biomass Bioenergy* **2018**, *111*, 154–164. [CrossRef]
3. Zhu, T.; Curtis, J.; Clancy, M. Promoting agricultural biogas and biomethane production: Lessons from cross-country studies. *Renew. Sustain. Energy Rev.* **2019**, *114*, 109332. [CrossRef]
4. Scarlat, N.; Dallemand, J.-F.; Fahl, F. Biogas: Developments and perspectives in Europe. *Renew. Energy* **2018**, *129*, 457–472. [CrossRef]
5. Barragán, E.A.; Ruiz, J.M.O.; Tigre, J.D.C.; Zalamea-León, E. Assessment of Power Generation Using Biogas from Landfills in an Equatorial Tropical Context. *Sustainability* **2020**, *12*, 2669. [CrossRef]
6. Persson, M.; Jonsson, O.; Wellinger, A. *Biogas Upgrading to Vehicle Fuel Standards and Grid*; IEA Bioenergy Task 37; Susanne AUER: Vienna, Austria, 2007.
7. Rincón, C.A.; De Guardia, A.; Couvert, A.; Wolbert, D.; Le Roux, S.; Soutrel, I.; Nunes, G. Odor concentration (OC) prediction based on odor activity values (OAVs) during composting of solid wastes and digestates. *Atmos. Environ.* **2019**, *201*, 1–12. [CrossRef]
8. Rasi, S. *Biogas Composition and Upgrading to Biomethane*; University of Jyväskylä: Jyväskylä, Finland, 2009; ISBN 978-951-39-3618-1.
9. Carrera-Chapela, F.; Donoso-Bravo, A.; Souto, J.A.; Ruiz-Filippi, G. Modeling the Odor Generation in WWTP: An Integrated Approach Review. *Water Air Soil Pollut.* **2014**, *225*, 1–15. [CrossRef]
10. Sevimoğlu, O.; Tansel, B. Effect of persistent trace compounds in landfill gas on engine performance during energy recovery: A case study. *Waste Manag.* **2013**, *33*, 74–80. [CrossRef]
11. Noorain, R.; Kindaichi, T.; Ozaki, N.; Aoi, Y.; Ohashi, A. Biogas purification performance of new water scrubber packed with sponge carriers. *J. Clean. Prod.* **2019**, *214*, 103–111. [CrossRef]
12. Vo, T.T.; Wall, D.M.; Ring, D.; Rajendran, K.; Murphy, J.D. Techno-economic analysis of biogas upgrading via amine scrubber, carbon capture and ex-situ methanation. *Appl. Energy* **2018**, *212*, 1191–1202. [CrossRef]
13. Rasi, S.; Läntelä, J.; Veijanen, A.; Rintala, J. Landfill gas upgrading with countercurrent water wash. *Waste Manag.* **2008**, *28*, 1528–1534. [CrossRef] [PubMed]
14. Rasi, S.; Läntelä, J.; Rintala, J. Upgrading landfill gas using a high pressure water absorption process. *Fuel* **2014**, *115*, 539–543. [CrossRef]
15. Devia, C.R.; Subrenat, A. Absorption of a linear (L2) and a cyclic (D4) siloxane using different oils: Application to biogas treatment. *Environ. Technol.* **2013**, *34*, 3117–3127. [CrossRef] [PubMed]
16. Ghorbel, L.; Tatin, R.; Couvert, A. Relevance of an organic solvent for absorption of siloxanes. *Environ. Technol.* **2013**, *35*, 372–382. [CrossRef]
17. Ajhar, M.; Travesset, M.; Yüce, S.; Melin, T. Siloxane removal from landfill and digester gas—A technology overview. *Bioresour. Technol.* **2010**, *101*, 2913–2923. [CrossRef]
18. Shen, M.; Zhang, Y.; Hu, D.; Fan, J.; Zeng, G. A review on removal of siloxanes from biogas: With a special focus on volatile methylsiloxanes. *Environ. Sci. Pollut. Res.* **2018**, *25*, 30847–30862. [CrossRef]
19. Schuur, B.; Brouwer, T.; Smink, D.; Sprakel, L.M. Green solvents for sustainable separation processes. *Curr. Opin. Green Sustain. Chem.* **2019**, *18*, 57–65. [CrossRef]
20. Zhang, S.; Zhang, X.; Dong, H.; Zhao, Z.; Zhang, S.; Huang, Y. Carbon capture with ionic liquids: Overview and progress. *Energy Environ. Sci.* **2012**, *5*, 6668–6681. [CrossRef]

21. Blanchard, L.A.; Hancu, D.; Beckman, E.J.; Brennecke, J.F. Green processing using ionic liquids and CO_2. *Nat. Cell Biol.* **1999**, *399*, 28–29. [CrossRef]
22. Reddy, R.G. Novel applications of ionic liquids in materials processing. *J. Phys. Conf. Ser.* **2009**, *165*, 1–6. [CrossRef]
23. Sarmad, S.; Mikkola, J.-P.; Ji, X. Carbon Dioxide Capture with Ionic Liquids and Deep Eutectic Solvents: A New Generation of Sorbents. *ChemSusChem* **2017**, *10*, 324–352. [CrossRef] [PubMed]
24. Makoś, P.; Fernandes, A.; Przyjazny, A.; Boczkaj, G. Sample preparation procedure using extraction and derivatization of carboxylic acids from aqueous samples by means of deep eutectic solvents for gas chromatographic-mass spectrometric analysis. *J. Chromatogr. A* **2018**, *1555*, 10–19. [CrossRef] [PubMed]
25. Makoś, P.; Przyjazny, A.; Boczkaj, G. Hydrophobic deep eutectic solvents as "green" extraction media for polycyclic aromatic hydrocarbons in aqueous samples. *J. Chromatogr. A* **2018**, *1570*, 28–37. [CrossRef] [PubMed]
26. Makoś, P.; Słupek, E.; Gębicki, J. Extractive detoxification of feedstocks for the production of biofuels using new hydrophobic deep eutectic solvents—Experimental and theoretical studies. *J. Mol. Liq.* **2020**, *308*, 113101. [CrossRef]
27. Makoś, P.; Słupek, E.; Gębicki, J. Hydrophobic deep eutectic solvents in microextraction techniques—A review. *Microchem. J.* **2020**, *152*, 104384. [CrossRef]
28. Słupek, E.; Makoś, P.; Gębicki, J. Deodorization of model biogas by means of novel non- ionic deep eutectic solvent. *Arch. Environ. Prot.* **2020**, *46*, 41–46. [CrossRef]
29. Słupek, E.; Makoś, P. Absorptive Desulfurization of Model Biogas Stream Using Choline Chloride-Based Deep Eutectic Solvents. *Sustainability* **2020**, *12*, 1619. [CrossRef]
30. Słupek, E.; Makos, P.; Dobrzyniewski, D.; Szulczynski, B.; Gebicki, J. Process control of biogas purification using electronic nose. *Chem. Eng. Trans.* **2020**, *82*. [CrossRef]
31. Shukla, S.K.; Mikkola, J.-P. Unusual temperature-promoted carbon dioxide capture in deep-eutectic solvents: The synergistic interactions. *Chem. Commun.* **2019**, *55*, 3939–3942. [CrossRef]
32. Trivedi, T.J.; Lee, J.H.; Lee, H.J.; Jeong, Y.K.; Choi, J.W. Deep eutectic solvents as attractive media for CO_2 capture. *Green Chem.* **2016**, *18*, 2834–2842. [CrossRef]
33. Shukla, S.K.; Mikkola, J.-P. Intermolecular interactions upon carbon dioxide capture in deep-eutectic solvents. *Phys. Chem. Chem. Phys.* **2018**, *20*, 24591–24601. [CrossRef] [PubMed]
34. Makoś, P.; Słupek, E.; Małachowska, A. Silica Gel Impregnated by Deep Eutectic Solvents for Adsorptive Removal of BTEX from Gas Streams. *Materials* **2020**, *13*, 1894. [CrossRef]
35. Słupek, E.; Makoś, P.; Gębicki, J. Theoretical and Economic Evaluation of Low-Cost Deep Eutectic Solvents for Effective Biogas Upgrading to Bio-Methane. *Energies* **2020**, *13*, 3379. [CrossRef]
36. Mu, T.; Rarey, J.; Gmehling, J. Performance of COSMO-RS with Sigma Profiles from Different Model Chemistries. *Ind. Eng. Chem. Res.* **2007**, *46*, 6612–6629. [CrossRef]
37. Klamt, A. Conductor-like Screening Model for Real Solvents: A New Approach to the Quantitative Calculation of Solvation Phenomena. *J. Phys. Chem.* **1995**, *99*, 2224–2235. [CrossRef]
38. Klamt, A. Prediction of the mutual solubilities of hydrocarbons and water with COSMO-RS. *Fluid Phase Equilib.* **2003**, *206*, 223–235. [CrossRef]
39. Klamt, A.; Eckert, F. COSMO-RS: A novel and efficient method for the a priori prediction of thermophysical data of liquids. *Fluid Phase Equilib.* **2000**, *172*, 43–72. [CrossRef]
40. Tansel, B.; Surita, S.C. Managing siloxanes in biogas-to-energy facilities: Economic comparison of pre- vs post-combustion practices. *Waste Manag.* **2019**, *96*, 121–127. [CrossRef]
41. Klamt, A. The COSMO and COSMO-RS solvation models. *Wiley Interdiscip. Rev. Comput. Mol. Sci.* **2018**, *8*, 1–11. [CrossRef]
42. Chu, Y.; He, X. MoDoop: An Automated Computational Approach for COSMO-RS Prediction of Biopolymer Solubilities in Ionic Liquids. *ACS Omega* **2019**, *4*, 2337–2343. [CrossRef]
43. Mullins, E.; Oldland, R.; Liu, Y.A.; Wang, S.; Sandler, S.I.; Chen, C.-C.; Zwolak, A.M.; Seavey, K.C. Sigma-Profile Database for Using COSMO-Based Thermodynamic Methods. *Ind. Eng. Chem. Res.* **2006**, *45*, 4389–4415. [CrossRef]
44. Jibril, B.E.-Y.; Mjalli, F.S.; Naser, J.; Gano, Z.S. New tetrapropylammonium bromide-based deep eutectic solvents: Synthesis and characterizations. *J. Mol. Liq.* **2014**, *199*, 462–469. [CrossRef]
45. García, G.; Aparicio, S.; Ullah, R.; Atilhan, M. Deep Eutectic Solvents: Physicochemical Properties and Gas Separation Applications. *Energy Fuels* **2015**, *29*, 2616–2644. [CrossRef]
46. Santiago, R.; Moya, C.; Palomar, J. Siloxanes capture by ionic liquids: Solvent selection and process evaluation. *Chem. Eng. J.* **2020**, *401*, 126078. [CrossRef]
47. Han, J.; Dai, C.; Yu, G.; Lei, Z. Parameterization of COSMO-RS model for ionic liquids. *Green Energy Environ.* **2018**, *3*, 247–265. [CrossRef]
48. Aissaoui, T. Neoteric FT-IR Investigation on the Functional Groups of Phosphonium- Based Deep Eutectic Solvents. *Pharm. Anal. Acta* **2015**, *6*, 10–12. [CrossRef]
49. Zhu, S.; Li, H.; Zhu, W.; Jiang, W.; Wang, C.; Wu, P.; Zhang, Q.; Li, H. Vibrational analysis and formation mechanism of typical deep eutectic solvents: An experimental and theoretical study. *J. Mol. Graph. Model.* **2016**, *68*, 158–175. [CrossRef]
50. Ghaedi, H.; Ayoub, M.; Sufian, S.; Lal, B.; Uemura, Y. Thermal stability and FT-IR analysis of Phosphonium-based deep eutectic solvents with different hydrogen bond donors. *J. Mol. Liq.* **2017**, *242*, 395–403. [CrossRef]

51. Shameli, K.; Bin Ahmad, M.; Jazayeri, S.D.; Sedaghat, S.; Shabanzadeh, P.; Jahangirian, H.; Shahri, M.M.; Abdollahi, Y. Synthesis and Characterization of Polyethylene Glycol Mediated Silver Nanoparticles by the Green Method. *Int. J. Mol. Sci.* **2012**, *13*, 6639–6650. [CrossRef]
52. Banjare, M.K.; Behera, K.; Satnami, M.L.; Pandey, S.; Ghosh, K.K. Self-assembly of a short-chain ionic liquid within deep eutectic solvents. *RSC Adv.* **2018**, *8*, 7969–7979. [CrossRef]
53. Maheswari, A.U.; Palanivelu, K. Carbon Dioxide Capture and Utilization by Alkanolamines in Deep Eutectic Solvent Medium. *Ind. Eng. Chem. Res.* **2015**, *54*, 11383–11392. [CrossRef]
54. Ruß, C.; König, B. Low melting mixtures in organic synthesis—An alternative to ionic liquids? *Green Chem.* **2012**, *14*, 2969–2982. [CrossRef]
55. Xydis, G.; Nanaki, E.A.; Koroneos, C.J. Exergy analysis of biogas production from a municipal solid waste landfill. *Sustain. Energy Technol. Assess.* **2013**, *4*, 20–28. [CrossRef]
56. Yusof, R.; Abdulmalek, E.; Sirat, K.; Rahman, M.B.A. Tetrabutylammonium Bromide (TBABr)-Based Deep Eutectic Solvents (DESs) and Their Physical Properties. *Molecules* **2014**, *19*, 8011–8026. [CrossRef]
57. Burrell, G.L.; Dunlop, N.F.; Separovic, F. Non-Newtonian viscous shear thinning in ionic liquids. *Soft Matter* **2010**, *6*, 2080–2086. [CrossRef]
58. Basaiahgari, A.; Panda, S.; Gardas, R.L. Acoustic, volumetric, transport, optical and rheological properties of Benzyltripropylammonium based Deep Eutectic Solvents. *Fluid Phase Equilib.* **2017**, *448*, 41–49. [CrossRef]
59. Verduzco, L.F.R. Density and viscosity of biodiesel as a function of temperature: Empirical models. *Renew. Sustain. Energy Rev.* **2013**, *19*, 652–665. [CrossRef]
60. Altamash, T.; Atilhan, M.; Aliyan, A.; Ullah, R.; Nasser, M.S.; Aparicio, S. Rheological, Thermodynamic, and Gas Solubility Properties of Phenylacetic Acid-Based Deep Eutectic Solvents. *Chem. Eng. Technol.* **2017**, *40*, 778–790. [CrossRef]
61. Sigma Aldrich. Safety Data Sheet Tetraethylene Glycol. Available online: https://www.sigmaaldrich.com/MSDS/MSDS/DisplayMSDSPage.do?country=PL&language=pl&productNumber=110175&brand=ALDRICH&PageToGoToURL=https%3A%2F%2Fwww.sigmaaldrich.com%2Fcatalog%2Fproduct%2Faldrich%2F110175%3Flang%3Dpl (accessed on 18 November 2020).
62. Abbott, A.P.; Boothby, D.; Capper, G.; Davies, D.L.; Rasheed, R.K. Deep Eutectic Solvents Formed between Choline Chloride and Carboxylic Acids: Versatile Alternatives to Ionic Liquids. *J. Am. Chem. Soc.* **2004**, *126*, 9142–9147. [CrossRef]
63. Hsu, C.H.; Chu, H.; Cho, C.M. Absorption and reaction kinetics of amines and ammonia solutions with carbon dioxide in flue gas. *J. Air Waste Manag. Assoc.* **2003**, *53*, 246–252. [CrossRef]
64. Yincheng, G.; Zhenqi, N.; Wenyi, L. Comparison of removal efficiencies of carbon dioxide between aqueous ammonia and Na-solution in a fine spray column. *Energy Procedia* **2011**, *4*, 512–518. [CrossRef]
65. Horikawa, M.; Rossi, F.; Gimenes, M.; Costa, C.M.; Da Silva, M. Chemical absorption of H2S for biogas purification. *Braz. J. Chem. Eng.* **2004**, *21*, 415–422. [CrossRef]
66. Ma, C.; Liu, C.; Lu, X.; Ji, X. Techno-economic analysis and performance comparison of aqueous deep eutectic solvent and other physical absorbents for biogas upgrading. *Appl. Energy* **2018**, *225*, 437–447. [CrossRef]
67. Lemus, J.; Bedia, J.; Moya, C.; Alonso-Morales, N.; Gilarranz, M.A.; Palomar, J.; Rodriguez, J.J. Ammonia capture from the gas phase by encapsulated ionic liquids (ENILs). *RSC Adv.* **2016**, *6*, 61650–61660. [CrossRef]
68. Ryckebosch, E.; Drouillon, M.; Vervaeren, H. Techniques for transformation of biogas to biomethane. *Biomass Bioenergy* **2011**, *35*, 1633–1645. [CrossRef]
69. Urasaki, N.; Wong, C. Separation of low molecular siloxanes for electronic application by liquid-liquid extraction. *IEEE Trans. Electron. Packag. Manuf.* **1999**, *22*, 295–298. [CrossRef]
70. Sun, S.; Niu, Y.; Sun, Z.; Xu, Q.; Wei, X. Solubility properties and spectral characterization of sulfur dioxide in ethylene glycol derivatives. *RSC Adv.* **2014**, *5*, 8706–8712. [CrossRef]
71. Sigma Aldrich. Tetraethylene Glycol. Available online: https://www.sigmaaldrich.com/catalog/product/aldrich/110175?lang=pl&region=PL (accessed on 11 December 2020).
72. Sigma Aldrich. Tetrapropylammonium Bromide. Available online: https://www.sigmaaldrich.com/catalog/product/aldrich/225568?lang=pl&region=PL (accessed on 11 December 2020).
73. Kolegaberlin Motor Oil. Available online: https://www.kolegaberlin.pl/product-pol-4300-Elf-Evolution-900-NF-5W40-5L.html (accessed on 30 November 2020).
74. Hoffmann Group. Cutting Oil. Available online: https://www.hoffmann-group.com/GB/en/houk/Power-tools-and-workshop-supplies/Cooling-lubricants/High-performance-cutting-oil-chlorine-free-Alpha-93/p/084210 (accessed on 30 November 2020).

Article

Behavior of Ternary Mixtures of Hydrogen Bond Acceptors and Donors in Terms of Band Gap Energies

Alberto Mannu [1,*], Francesca Cardano [1], Salvatore Baldino [1] and Andrea Fin [2]

1. Department of Chemistry, University of Turin, Via Pietro Giuria 7, I-10125 Turin, Italy; francescacardano@gmail.com (F.C.); salvatore.baldino@unito.it (S.B.)
2. Department of Drug Science and Technology, University of Turin, Via Pietro Giuria 9, I-10125 Turin, Italy; andrea.fin@unito.it
* Correspondence: albertomannu@gmail.com

Abstract: Three ternary mixtures composed by choline chloride (ChCl), ethylene glycol (EG), and a second hydrogen bond donor (HBD) as ethanol (A), 2-propanol (B), and glycerol (C) were studied in terms of composition related to the band gap energy (BGE). A Design of Experiments (DoE) approach, and in particular a *Simple Lattice* three-components design, was employed for determining the variation of the BGE upon the composition of each system. UV-VIS analysis and subsequent Tauc plot methodology provided the data requested from the DoE, and multivariate statistical analysis revealed a drop of the BGE in correspondence to specific binary compositions for systems A and B. In particular, a BGE of 3.85 eV was registered for the mixtures ChCl/EtOH (1:1) and ChCl/2-propanol (1:1), which represents one of the lowest values ever observed for these systems.

Keywords: eutectic mixtures; hydrogen bond acceptor; hydrogen bond donor; design of experiments; Tauc plot; band gap energy

1. Introduction

Hydrogen bond-based systems have been affirmed during the last 20 years as one of the most recurrent topics in the scientific literature [1]. In particular, the possibility to combine in a eutectic molar ratio Hydrogen Bond Acceptors (HBAs) and Hydrogen Bond Donors (HBDs) and form liquid mixtures at room temperature with increased solvent ability, which was reported for the first time by Abbott et al. in 2003 (choline chloride/urea 1:2) [2], created opportunity for many applications in several research and industrial sectors. Since the first paper of Abbot about the topic, dozens of such systems have been developed, characterized, and studied in terms of physical–chemical properties [3–5]. These systems are of particular interest when the molar ratio between HBAs and HBDs produces a drop of the melting point which produces experimental results deeper than the expected theoretical ones [3,6]. When this specific combination takes place, a concomitant increased solvent ability is also observed [7], and for this reason, the acronym DESs has been proposed, which stands for Deep Eutectic Solvents. Through the years, many researchers have been triggered by the possibility of engineering DESs by choosing opportune combinations of HBAs and HBDs and by finding the best molar ratio between them [6]. At the molecular level, most of the DESs can be described according to the hole theory [8] and rationalized as systems made by an intense hydrogen bond network decorated with randomly distributed holes, where the ions can move along the network by jumping from one hole to another [9]. This supramolecular behavior gives to the system peculiar properties such as an increased density, a decreased viscosity, and a low conductivity [1]. On the basis of such characteristics, DESs have found many applications as media for biomass treatment [10], metal extraction [11], solvents for Volatile Organic Compounds (VOCs) [12,13], templates for ionothermal synthesis [14–16], or non-innocent solvents in organic synthesis [17–21], as well as additives in pharmaceutical formulations [1,22,23].

Recently, some preliminary studies have highlighted how the deep eutectic composition in mixtures of hydrogen bond acceptors and donors is related to a depression of the band gap energy [24,25] along with a drop of the structural disorder (Urbach energy) [26]. The possibility to tune such optical parameters results is of interest especially for the development of new liquid organic semiconductors.

In this context, the possibility to model the band gap energy (BGE) and monitoring its variation in ternary mixtures composed by one HBA (choline chloride, ChCl), a first HBD (ethylene glycol, EG), and a second HBD (ethanol-EtOH (A), 2-propanol (B), or glycerol-GLY (C)) is herein explored. In particular, a Design of Experiments (DoE) approach followed by multivariate analysis was employed to finally plot a surface representative of the variation of the BGE depending on the molar fraction ratio between the constituents of each system. According to the DoE, seven samples were prepared for each one of the three systems (A, B, C) for a total of 21 experiments, which were subjected to the graphical Tauc plot method for the determination of the BGE.

2. Experimental Section

2.1. General Synthetic Procedure

Chemicals were purchased from commercial sources and used as received. In particular, choline chloride (>98%) and 2-propanol (99.8%) were purchased from Merck KGaA, Darmstadt, Germany, ethanol (96%) and glycerol (99.6%) were purchased from VWR, and ethylene glycol (99%) was purchased from Carlo Erba. Finally, H_2O was purified with a Millipore RiO$_s$ 3 Water System.

The samples were prepared following this protocol: ChCl was weighted in a vial, and 10 wt % of water was added. Thus, one or two hydrogen bond donors were added, and the resulting mixture was stirred for 2 h at room temperature before the analysis.

2.2. Spectroscopic UV-VIS Analysis

The samples were analyzed in a pure form by UV-VIS spectrophotometry. The spectra were recorded in transmittance mode in a quartz cell (path length: 1.00 mm) with an Agilent Cary 60 UV-Vis Spectrophotometer.

2.3. Statistical Analysis

Design of Experiment (DoE): a *Simple Lattice* three-components design was settled up [27] Thus, for each ternary system, we prepared seven samples with the molar ratio as reported in Table 1.

Table 1. BGEs and corresponding nomenclature for the systems A–C.

HBA (χ) [1]	HBD (χ)	HBD (χ)	Name	BGE (eV)
ChCl (1)	EG (0)	EtOH (0)	A1	5.75
ChCl (0)	EG (1)	EtOH (0)	A2	5.66
ChCl (0)	EG (0)	EtOH (1)	A3	5.88
ChCl (0.5)	EG (0.5)	EtOH (0)	A4	5.86
ChCl (0.5)	EG (0)	EtOH (0.5)	A5	3.85
ChCl (0)	EG (0.5)	EtOH (0.5)	A6	6.05
ChCl (0.33)	EG (0.33)	EtOH (0.33)	A7	5.87
ChCl (1)	EG (0)	2-propanol (0)	B1	5.75
ChCl (0)	EG (1)	2-propanol (0)	B2	5.66
ChCl (0)	EG (0)	2-propanol (1)	B3	5.88
ChCl (0.5)	EG (0.5)	2-propanol (0)	B4	5.87
ChCl (0.5)	EG (0)	2-propanol (0.5)	B5	3.85
ChCl (0)	EG (0.5)	2-propanol (0.5)	B6	6.05
ChCl (0.33)	EG (0.33)	2-propanol (0.33)	B7	5.84
ChCl (1)	EG (0)	GLY (0)	C1	5.75

Table 1. *Cont.*

HBA (χ) [1]	HBD (χ)	HBD (χ)	Name	BGE (eV)
ChCl (0)	EG (1)	GLY (0)	C2	5.66
ChCl (0)	EG (0)	GLY (1)	C3	5.23
ChCl (0.5)	EG (0.5)	GLY (0)	C4	5.86
ChCl (0.5)	EG (0)	GLY (0.5)	C5	5.51
ChCl (0)	EG (0.5)	GLY (0.5)	C6	5.17
ChCl (0.33)	EG (0.33)	GLY (0.33)	C7	5.12

[1] To each ChCl sample 10 wt % of water was added to make the samples measurable at UV-VIS.

Multivariate analysis, including the Analysis of the Variance (ANOVA) and the corresponding surface plots, was conducted with the Statgraphics Centurion v 18 software, Statgraphics Technologies, Inc. The Plains, Virginia.

3. Results and Discussion

Ternary systems A, B, and C were prepared by combining ChCl, EG, and a second HBD. The choice of the second HBD was driven by the affinity of alcohols with ChCl and EG; thus, EtOH, 2-propanol, and GLY were selected.

The aim of the research was to develop a tool for engineering a ternary mixture of HBAs and HBDs in terms of BGE. Thus, the first target was to model the variation of the BGE in each ternary system A, B, and C and to provide a suitable statistical instrument for describe these mixtures. In order to reach such goal, a DoE approach was used, and in particular, a *Simple Lattice* three-components mixture experiment was implemented [28].

Seven different molar combinations were prepared for each system, and the corresponding BGEs were determined by the known graphic UV-VIS-based Tauc plot method, following a procedure previous optimized by some of us [24].

In Table 1, the nomenclature corresponding to the systems prepared and subjected to UV-VIS analysis is reported.

In Figure 1, the UV-VIS spectra of representative binary systems A5, C5, and the thernary ones A7, C7 are reported.

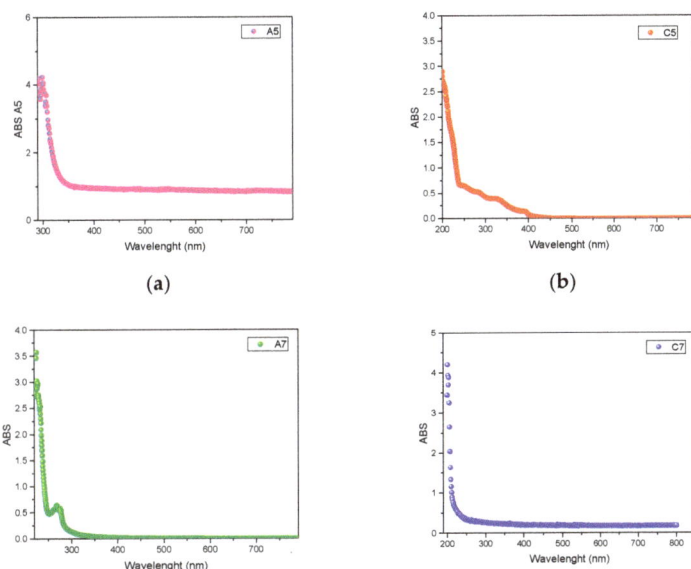

Figure 1. UV-VIS spectra of systems A5 (**a**), C5 (**b**), A7 (**c**), and C7 (**d**).

From a visual and qualitative analysis of the UV-VIS spectra of the systems reported in Figure 1, it is possible to notice some differences. In particular, systems A5 (a, ChCl/EG 1/1), and C7 (d, ChCl/EG/GLY 1/1/1) show the same UV-VIS behavior, suggesting that the addition of glycerol does not affect in a relevant way the optical characteristics of the system. On the other side, systems C5 (b, ChCl/GLY 1/1) and A7 (c, ChCl/EG/EtOH 1/1/1) reveal different absorbance spectra, with C5 showing multiple absorbance peaks between 200 and 400 nm, and A7 highlighting a relevant peak at 270 nm. Nevertheless, it is difficult to extrapolate trends and information from the analysis of the UV-VIS spectra. On the other side, the calculation of the band gap energy from the UV-VIS data can provide many information, which can be related with the structural effects of the constituents on the system.

Looking at the band gap data reported in Table 1, it is possible to highlight some trends. Each single component, measured in pure form, shows a relatively high BGE: BGE_{ChCl} 5.75 eV, BGE_{EtOH} 5.88 eV, BGE_{GLY} 5.23 eV, and BGE_{EG} 5.66 eV. As expected, when two HBDs are combined (1:1 molar ratio), no significant drop of the BGE is observed: $BGE_{EG/EtOH}$ 6.05 eV (A6), $BGE_{EG/2\text{-}propanol}$ 6.05 (B6), $BGE_{EG/GLY}$ 5.17 eV (C6). In addition, no relevant differences were observed changing EtOH by 2-propanol (A6 vs. B6). As a matter of fact, the deepest reduction of the BGE was observed for the binary systems A5 and B5, which do not contain EG. This value of BGE, around 3.86 eV, falls in the range of interest for potential application as an organic liquid semiconductor. It is interesting to notice that the corresponding binary system C5, composed by ChCl and GLY, shows a BGA far away from the parent A5 and B5.

In Figure 2, the reduction of the BGE obtained by substituting EG with EtOH (A4, B4, and C4 vs. A5) is reported.

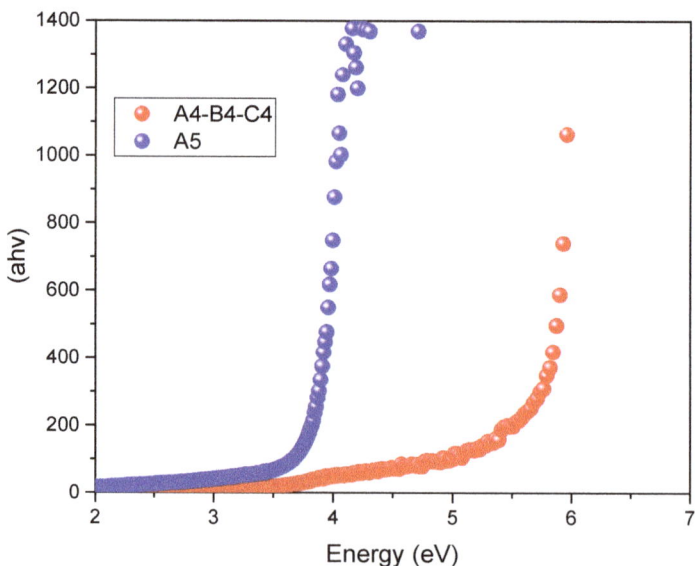

Figure 2. Tauc plots relative to the systems A4, B4, C4, ChCl/EG (1:1) and A5, ChCl/EtOH (1:1).

Each set of experiments (A, B, and C, Table 1) was processed according to the DoE procedure adopted to build a descriptive model of the variation of the BGE as a function of the molar ratio between the three constituents of the mixture.

Multivariate Analysis of Systems A, B, and C

At first, system A composed by ChCl/EG/EtOH was analyzed with the aim to find the best statistical model that can represent the behavior of the mixture in terms of variation of the BGE. After a screening between linear, special cubic, and quadratic statistical models, the last one has been selected as the most accurate in describing the system. Full data details are reported in the Supporting Information file.

Starting from the selected quadratic model, the data obtained for the system A were subjected to the Analysis of the Variance (ANOVA), which gave the results reported in Table 2.

Table 2. ANOVA for BGE of system A.

Source	Sum of Squares	Df	Mean Square	F-Ratio	*p*-Value
Quadratic model	3.1813	5	0.63626	1.48	0.5435
Total error	0.430699	1	0.430699		
Total (corr.)	3.612	6			

ANOVA analysis shows an R-squared value that indicates that the model as fitted explains 88.0759% of the variability in BGE. The adjusted R-squared statistic, which is more suitable for comparing models with different numbers of independent variables, is 28.4553%. The standard error of the estimate shows the standard deviation of the residuals to be 0.656277. The mean absolute error (MAE) of 0.195845 is the average value of the residuals. The Durbin–Watson (DW) statistic tests the residuals to determine if there is any significant correlation based on the order in which they occur in your data file. Since the *p*-value is greater than 5.0%, there is no indication of serial autocorrelation in the residuals at the 5.0% significance level.

Systems B and C were subjected to the same statistical treatment confirming the quadratic model as the best one.

For comparison purposes, the R-squared values of systems A, B, and C are reported in Table 3.

Table 3. R-squared for system A, B, and C.

System	Statistical Model	R-Squared (%)
A	quadratic	88.0759
B	quadratic	89.2872
C	quadratic	83.3460

Once we determined the statistical parameters that better described the behavior of each system, it is possible to graphically represent them in the form of a responsive surface (Figure 3).

The surface responding plots reported in Figure 3 describe the variation of the systems A-C in terms of BGE. From a fist qualitative analysis, it is possible to notice that the shape of the surface that describes the behavior of the ternary mixture is very similar for systems A and B, while it changes for system C. This is mainly due to the previously commented different interaction between ChCl and GLY (C5) with respect to ChCl and EtOH (A5) or ChCl and 2-propanol (B5). This experimental behavior of C5, combined with lower maximum values of BGE for A6 and B6 (6.04 eV), determines a flatter surface. From the combined analysis of the plots reported in Figure 3, it is possible to conclude that only systems A and C show a consistent depression of the BGE, which indeed correspond to a binary system. No one of the three systems considered performs better (in terms of lower BGE) with a ternary composition. Thus, the increment of O-H bonds achieved with the introduction of a second HBD seems to negatively affect the eutectic nature of the mixture.

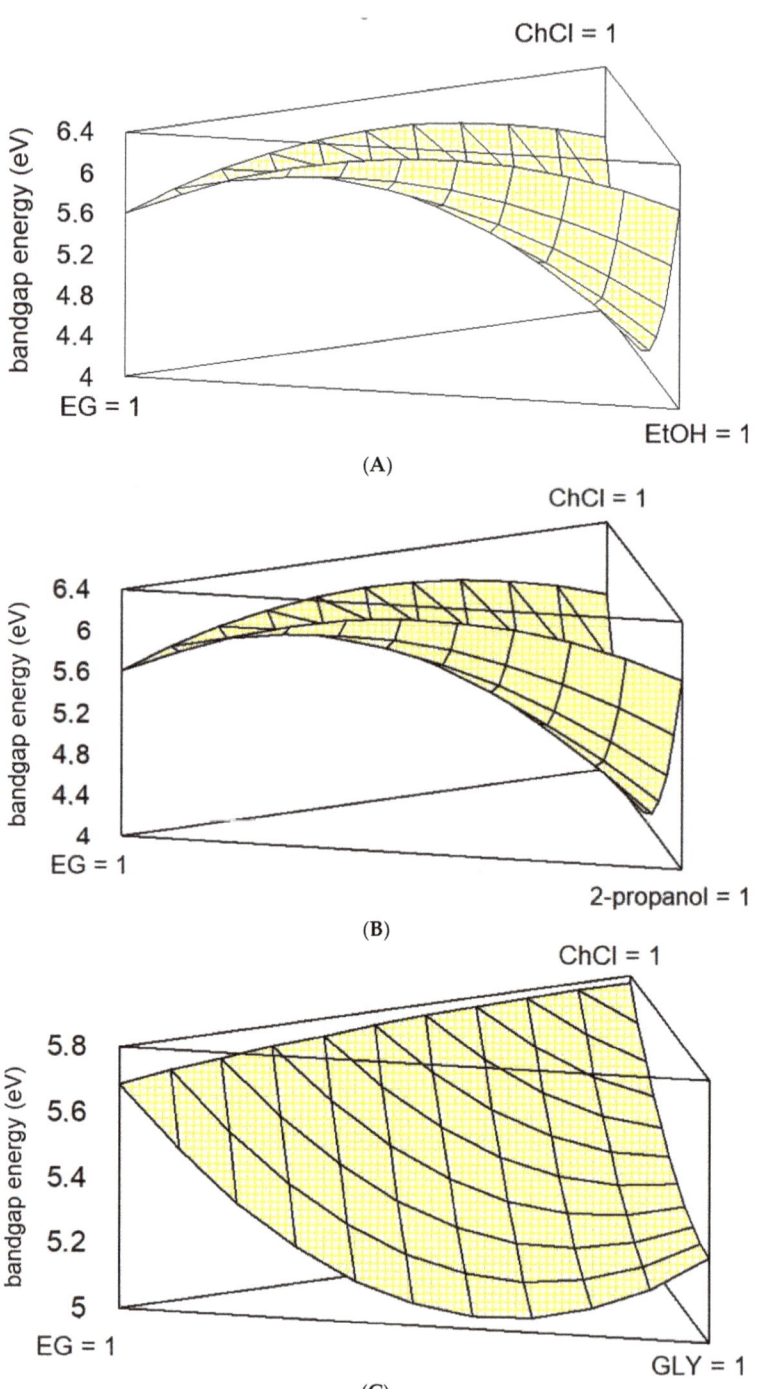

Figure 3. Estimated response surface for systems (**A**–**C**).

To the best of our knowledge, this study represents the first report about the variation of the BGE in ternary mixtures of HBAs and HBDs. In addition, the statistical model herein presented can be applied for optimizing other systems, even considering different parameters beyond the BGE.

4. Conclusions

Three ternary systems composed by ChCl, EG, and a second HBD (EtOH, 2-propanol, GLY) were studied in terms of variation of the BGE with respect to the molar ration of the former's components. A statistical reliable model that describes the relationship between BGE and molar composition was built and described. The statistical multivariate analysis revealed, for the systems herein considered, that an excessive increasing of the O-H bonds affects the eutectic nature of the mixture, resulting in an increasing of the BGE. In addition, the combination between UV-VIS spectroscopy, Tauc plot method (for the band gap energy determination), and the Simple Lattice DoE followed by statistical multivariate analysis provide an easy and fast tool for engineering ternary mixtures of Hydrogen Bond Donors (HBDs) and Acceptors (HBAs). In fact, the combination of techniques reported allow mapping the variation of the band gap energy versus the molar composition of the ternary system. This procedure can be used for screening purposes with the target to select the best combination between HBDs and HBAs that provides the minimum band gap energy value.

Author Contributions: Conceptualization, A.M., S.B. and A.F.; methodology, A.M. and A.F.; software, S.B.; validation, A.M.; formal analysis, F.C.; investigation, A.M.; resources, A.M.; data curation, F.C.; writing—original draft preparation, A.M.; writing—review and editing, F.C., S.B. and A.F.; supervision, A.M.; funding acquisition, A.M. All authors have read and agreed to the published version of the manuscript.

Funding: This research received no external funding.

Institutional Review Board Statement: Not applicable.

Informed Consent Statement: Not applicable.

Data Availability Statement: External data are not available.

Conflicts of Interest: The authors declare no conflict of interest.

References

1. Mannu, A.; Blangetti, M.; Baldino, S.; Prandi, C. Promising Technological and Industrial Applications of Deep Eutectic Systems. *Materials* **2021**, *14*, 2494. [CrossRef] [PubMed]
2. Abbott, A.P.; Capper, G.; Davies, D.L.; Rasheed, R.K.; Tambyrajah, V. Novel solvent properties of choline chloride/urea mixtures. *Chem. Commun.* **2003**, *1*, 70–71. [CrossRef] [PubMed]
3. Smith, E.L.; Abbott, A.P.; Ryder, K.S. Deep eutectic solvents (DESs) and their applications. *Chem. Rev.* **2014**, *114*, 11060–11082. [CrossRef]
4. Florindo, C.; Branco, L.C.; Marrucho, I.M. Quest for green-solvent design: From hydrophilic to hydrophobic (deep) eutectic solvents. *ChemSusChem* **2019**, *12*, 1549–1559. [CrossRef]
5. Liu, Y.; Friesen, J.B.; McAlpine, J.B.; Lankin, D.C.; Chen, S.-N.; Pauli, G.F.J. Natural deep eutectic solvents: Properties, applications, and perspectives. *Nat. Prod.* **2018**, *81*, 679–690. [CrossRef]
6. Martins, M.A.R.; Pinho, S.P.; Coutinho, J.A.P.J. Insights into the nature of eutectic and deep eutectic mixtures. *Solut. Chem.* **2019**, *48*, 962–982. [CrossRef]
7. Perna, F.M.; Vitale, P.; Capriati, V. Deep eutectic solvents and their applications as green solvents. *Curr. Opin. Green Sustain. Chem.* **2020**, *21*, 27–33. [CrossRef]
8. Abbott, A.P. Application of Hole Theory to the Viscosity of Ionic and Molecular Liquids. *ChemPhysChem* **2004**, *5*, 1242–1246. [CrossRef]
9. Abbott, A.P.; Capper, G.; Gray, S. Design of Improved Deep Eutectic Solvents Using Hole Theory. *ChemPhysChem* **2006**, *7*, 803–806. [CrossRef]
10. Li, X.; Row, K.H. Development of deep eutectic solvents applied in extraction and separation. *J. Sep. Sci.* **2016**, *39*, 3505–3520. [CrossRef]
11. Liu, P.; Hao, J.W.; Mo, L.P.; Zhang, Z.H. Recent advances in the application of deep eutectic solvents as sustainable media as well as catalysts in organic reactions. *RSC Adv.* **2015**, *5*, 48675–48704. [CrossRef]

12. Moura, L.; Moufawad, T.; Ferreira, M.; Bricout, H.; Tilloy, S.; Monflier, E.; Costa Gomes, M.F.; Landy, D.; Fourmentin, S. First evidence of cyclodextrin inclusion complexes in a deep eutectic solvent. *Environ. Chem. Lett.* **2017**, *15*, 747–753. [CrossRef]
13. Di Pietro, M.E.; Colombo Dugoni, G.; Ferro, M.; Mannu, A.; Castiglione, F.; Costa Gomes, M.; Fourmentin, S.; Mele, A. Do cyclodextrins encapsulate volatiles in deep eutectic systems? *ACS Sust. Chem Eng.* **2019**, *7*, 17397–17405. [CrossRef]
14. Maschita, J.; Banerjee, T.; Savasci, G.; Haase, F.; Ochsenfeld, C.; Lotsch, B.V. Ionothermal Synthesis of Imide-Linked Covalent Organic Frameworks. *Angew. Chem. Int. Ed.* **2020**, *59*, 15750–15758. [CrossRef]
15. Wu, J.; Wang, Y.; Zhang, Y.; Meng, H.; Xu, Y.; Han, Y.; Wang, Z.; Dong, Y.; Zhang, X.J. Highly safe and ionothermal synthesis of Ti3C2 MXene with expanded interlayer spacing for enhanced lithium storage. *Energy Chem.* **2020**, *47*, 203–209. [CrossRef]
16. Zhao, X.; Duan, W.; Wang, Q.; Ji, D.; Zhao, Y.; Li, G. Microwave-assisted ionothermal synthesis of Fe-LEV molecular sieve with high iron content in low-dosage of eutectic mixture. *Microporous Mesoporous Mater.* **2019**, *275*, 253–262. [CrossRef]
17. Qin, H.; Hu, X.; Wang, J.; Cheng, H.; Chen, L.; Qi, Z. Overview of acidic deep eutectic solvents on synthesis, properties and applications. *Green Energy Environ.* **2020**, *5*, 8–21. [CrossRef]
18. Nejrotti, S.; Mannu, A.; Blangetti, M.; Baldino, S.; Fin, A.; Prandi, C. Optimization of Nazarov Cyclization of 2, 4-Dimethyl-1, 5-diphenylpenta-1, 4-dien-3-one in Deep Eutectic Solvents by a Design of Experiments Approach. *Molecules* **2020**, *25*, 5726. [CrossRef]
19. Sanyal, U.; Yuk, S.F.; Koh, K.; Lee, M.-S.; Stoerzinger, K.; Zhang, D.; Meyer, L.C.; Lopez-Ruiz, J.A.; Karkamkar, A.; Holladay, J.D.; et al. Hydrogen bonding enhances the electrochemical hydrogenation of benzaldehyde in the aqueous phase. *Angew. Chem. Int. Ed.* **2021**, *60*, 290–296. [CrossRef]
20. Hooshmand, S.E.; Afshari, R.; Ramón, D.J.; Varma, R.S. Deep eutectic solvents: Cutting-edge applications in cross-coupling reactions. *Green Chem.* **2020**, *22*, 3668–3692. [CrossRef]
21. Cavallo, M.; Arnodo, D.; Mannu, A.; Blangetti, M.; Prandi, C.; Baratta, W.; Baldino, S. Deep eutectic solvents as H2-sources for Ru (II)-catalyzed transfer hydrogenation of carbonyl compounds under mild conditions. *Tetrahedron* **2021**, *83*, 131997. [CrossRef]
22. Jablonský, M.; Škulcová, A.; Šima, J. Use of deep eutectic solvents in polymer chemistry—A review. *Molecules* **2019**, *24*, 3978. [CrossRef] [PubMed]
23. Duarte, A.R.C.; Ferreira, A.S.D.; Barreiros, S.; Cabrita, E.; Reis, R.L.; Paiva, A. A comparison between pure active pharmaceutical ingredients and therapeutic deep eutectic solvents: Solubility and permeability studies. *Eur. J. Pharm. Biopharm.* **2017**, *114*, 296–304. [CrossRef] [PubMed]
24. Mannu, A.; Ferro, M.; Colombo Dugoni, G.; Di Pietro, M.E.; Garroni, S.; Mele, A.J. From deep eutectic solvents to deep band gap systems. *Mol. Liq.* **2020**, *301*, 112441. [CrossRef]
25. Mannu, A.; Di Pietro, M.E.; Mele, A. Band-Gap energies of choline chloride and triphenylmethylphosphoniumbromide-based systems. *Molecules* **2020**, *25*, 1495. [CrossRef] [PubMed]
26. Mannu, A.; Cardano, F.; Fin, A.; Baldino, S.; Prandi, C.J. Choline chloride-based ternary deep band gap systems. *Mol. Liq.* **2021**, *330*, 115717. [CrossRef]
27. Lambrakis, D.P. Experiments with mixtures: A generalization of the simplex-lattice design. *J. R. Stat. Soc. Ser. B Methodol.* **1968**, *30*, 123–136. [CrossRef]
28. Fang, K.-T.; Liu, M.-Q.; Qin, H.; Zhou, Y.-D. *Theory and Application of Uniform Experimental Designs*; Springer: Berlin/Heidelberg, Germany, 2018; ISBN 978-981-13-2041-5.

Article

Effective and Selective Extraction of Quercetin from Onion (*Allium cepa* L.) Skin Waste Using Water Dilutions of Acid-Based Deep Eutectic Solvents

Matteo Ciardi [1,†], Federica Ianni [2,†], Roccaldo Sardella [2,3], Stefano Di Bona [1], Lina Cossignani [2,3], Raimondo Germani [1], Matteo Tiecco [1,*] and Catia Clementi [1]

1. Department of Chemistry, Biology and Biotecnology, University of Perugia, Via Elce di Sotto 8, 06123 Perugia, Italy; matteo.ciardi95@gmail.com (M.C.); dibonastefano@gmail.com (S.D.B.); raimondo.germani@unipg.it (R.G.); catia.clementi@unipg.it (C.C.)
2. Department of Pharmaceutical Sciences, University of Perugia, Via Fabretti 48, 06123 Perugia, Italy; federica.ianni@unipg.it (F.I.); roccaldo.sardella@unipg.it (R.S.); lina.cossignani@unipg.it (L.C.)
3. Center for Perinatal and Reproductive Medicine, University of Perugia, Santa Maria della Misericordia University Hospital, 06132 Perugia, Italy
* Correspondence: matteotiecco@gmail.com
† These authors contributed equally to this paper.

Abstract: Deep Eutectic Solvents (DESs) are experiencing growing interest as substitutes of polluting organic solvents for their low or absent toxicity and volatility. Moreover, they can be formed with natural bioavailable and biodegradable molecules; they are synthesized in absence of hazardous solvents. DESs are, inter alia, successfully used for the extraction/preconcentration of biofunctional molecules from complex vegetal matrices. Onion skin is a highly abundant waste material which represents a reservoir of molecules endowed with valuable biological properties such as quercetin and its glycosylated forms. An efficient extraction of these molecules from dry onion skin from "Dorata di Parma" cultivar was obtained with water dilution of acid-based DESs. Glycolic acid (with betaine 2/1 molar ratio and L-Proline 3/1 molar ratio as counterparts) and of *p*-toluensulphonic acid (with benzyltrimethylammonium methanesulfonate 1/1 molar ratio)-based DESs exhibited more than 3-fold higher extraction efficiency than methanol (14.79 µg/mL, 18.56 µg/mL, 14.83 µg/mL vs. 5.84 µg/mL, respectively). The extracted quercetin was also recovered efficaciously (81% of recovery) from the original extraction mixture. The proposed extraction protocol revealed to be green, efficacious and selective for the extraction of quercetin from onion skin and it could be useful for the development of other extraction procedures from other biological matrixes.

Keywords: quercetin; Deep Eutectic Solvents; extraction; preconcentration; DESs water dilutions; RP-HPLC-UV; UHPLC-MS/MS; recovery

1. Introduction

The substitution of common toxic and volatile organic solvents with novel greener liquids is of prior importance to tackle the urgent problems of planet pollution and improper chemical wastes disposal [1–3]. A large number of tons per year of volatile, toxic and bioaccumulating organic solvents are in fact used in chemical industries, playing the greatest part in chemical applications [4].

Deep Eutectic Solvents (DESs) are a novel class of organic liquids that are gaining increasing attraction in many sub-fields of chemical practice, as is well-documented by the recent literature [5–8]. A DES is an organic liquid endowed with valuable "green properties" which is formed via weak interactions, such as hydrogen bonds, between two (often solid) molecules, namely a hydrogen bond donor (HBD) and a hydrogen bond acceptor (HBA). The network of weak interactions established between the molecules of the same species and between the molecules of different species determines a difficult or even impossible crystal lattice formation. This ultimately leads to the formation of a liquid [9,10]. The resulting liquids show a deviation from ideal liquid mixtures in terms of the melting temperatures depending on the molar fraction of the components, with a deepening of the melting points as well as a shift in the molar ratio of the eutectic point [11]. Relevant papers are reported in literature with a quantitative approach to the phase diagrams of these liquids that can define these liquids as DESs or simple eutectic mixtures [12–14]. The preparation of these green mixtures represents a great step ahead in the formulation of innovative green liquids, especially whenever they are compared with other green liquids [15–17]. This is because the liquids are formed simply by heating and mixing the two often solid substances until homogenous systems are formed (often in few minutes) without the use of any other solvent. As a result, the realization processes, which often require only a few minutes, have 100% yield and 100% atom economy [5]. Many different molecules can be used to realize DESs liquids, including, but not limited to: onium salts with metal chlorides (also hydrated); choline chloride mixed with hydrogen bond donors (such as carboxylic acids); Lewis bases mixed with alcohols or amides; etc. [17–20].

The green properties of DESs rely on the fact that they generally (i) are non-toxic, (ii) have low or absent volatility (leading to the possibility of "out of the hood" procedures), (iii) are biodegradable and, (iv) in the case of natural source molecules used for their preparation (NADESs: Natural Deep Eutectic Solvents), their impact on the environment is markedly reduced, both in terms of bioavailability and biodegradability of the liquids themselves [21–25]. However, because the properties of the liquids' components are retained in the properties of the DESs, not all these novel liquids can be considered of course as totally green as they can be formed by harmful or not green molecules. This is the case i.e., for highly acidic DESs components as p-toluenesulfonic acid. In these cases, the use of these liquids can nevertheless permit the avoidance of volatile mineral acidic components [26].

For the same reasons of the properties of the constituting components, DESs can also exhibit appreciable catalytic properties [8,27–32]. The effect seems to depend on the "availability" of the molecule forming the DES in exerting the catalytic action [33–35].

Another interesting facet about DESs, which is steadily gaining ground in the literature, is represented by the water dilutions of these weak forces-based systems, as water molecules can participate in the network of weak interaction [36,37]. The increase of the water dilutions leads to a solvation of clusters of couples of HBD-HBA molecules and to micro-domains of DESs and water; at values over about 50–60% w/w the DESs' deconstruction occurs. Water dilutions, even after low amounts of added water, have peculiar and interesting physical-chemical properties as they show a high decrease of their viscosity and changes in their polarity [38]. These effects are however dependent on the hydrophobicity and hydrophilicity of the DESs' components, so the values of water needed to determine structural changes can slightly shift.

Based on the last property, DESs are fruitfully applied in extraction and preconcentration procedures from different matrices, encompassing vegetal and biological ones [39,40]. One of the most interesting areas where DESs are finding relevant results is in their use as green liquids for biomass feedstocks treatments, where they are finding high effectiveness [41–45].

In particular, phenolic compounds of vegetal origin are successfully extracted/preconcentrated as they can participate in the hydrogen bonds network with the hydroxyl function as well as with the aromatic portions that can act as hydrogen bond acceptors [24,46].

Among the several naturally occurring flavonoids, quercetin is currently one of the most extensively studied because, besides its proven anti-blood clotting, cardioprotective, neuroprotective, anti-inflammatory, anti-cancer and antioxidant properties [47,48], it also shows an effective antiviral and immunomodulatory activity. In particular, recent studies suggest the efficacy of quercetin-based formulations in reducing symptoms severity and negative predictors of severe acute respiratory syndrome coronavirus 2 (SARS-CoV-2), which is the cause of the present COVID-19 global pandemic [49,50]. However, further studies are still needed to demonstrate these relevant data [51].

Among the different methods available for the extraction of quercetin, and of flavonoids from natural sources more generally, the ones based on ultrasound- and microwave-assisted procedures are the most widely applied. Besides the major well-recognized advantages of these techniques (fast execution, a certain level of environmentally friendly character, easy automation, etc.), some of their severe limitations and major drawbacks have been also described [52]. For example, thermal degradation of the compounds of interest, as well as their accidental participation in unwanted/uncontrolled side reactions are worthy of noting. Still, the use of even small percentages of volatile organic extraction solvents can represent a problem in terms of their environmental and safety impact. Equally important, using ultrasound- and microwave-assisted procedures implies a proper tuning and combination of several process variables, which must be cautiously optimized.

Scientific works describing the use of DESs and their water dilutions for the effective quercetin extraction from onion skin waste are already present in the literature [53]; the authors reported the use of common DESs such as choline chloride/urea/water mixtures or sugar-based DESs. Furthermore, applications regarding the efficient extraction performances of high water dilutions of DESs have been described [54].

Quercetin is contained in abundance in different varieties of vegetables and fruits such as apples, honey, raspberries, onions, red grapes, cherries, citrus fruits, and green and red leafy vegetables [55]. Among them, high quercetin content is found in yellow onion skin. Onion is one of the most important horticultural crops, which has reached a current worldwide production of around 100 million tons in 2019 leading to a consequent generation of a consistent amount of solid waste material. Recent literature reports that the annual European production of onion waste is around 500,000 tons, especially in major producing countries such as Spain, the Netherlands and the United Kingdom [56]. Onion skin, the most highly abundant waste material derived from onion processing, represents a reservoir of molecules endowed with valuable biofunctional properties [57,58]. Within the (phyto)complex, quercetin and in its glycosylated forms occupy a prominent position in this regard [59,60].

In this work, we present an effective and green procedure for the extraction of quercetin and its principal glycosylated form from dry onion skin of "Dorata di Parma" cultivar with the use of water dilutions of acidic DESs with a heating/stirring- and ultrasound-assisted protocol. This procedure revealed to be much more effective than the use of neat methanol, a protic highly toxic and volatile solvent commonly used in the extractions of polyphenols from complex matrixes [61,62]. An anti-solvent and a reversed-phase chromatography approach were performed to enable the raw quercetin recovery from each extract. In order to select the best parameters to maximize the recovery of quercetin(s) from onion extracts, the extraction efficiency was monitored by reversed phase-high performance liquid chromatography coupled to UV detection (RP-HPLC-UV). Ultra-high performance liquid chromatography-tandem mass spectrometry (UHPLC-MS/MS) analyses were also performed to allow the identification of the main peaks.

2. Materials and Methods
2.1. Reagents and Instruments

Glycolic acid (GA), Trimethylglycine (TMG), Ethylene Glycol (EG), Choline Chloride (ChCl), Glycerol (GLY), Urea (U), *p*-toluenesulfonic acid (pTSA), L-proline (L-PRO), Octanoic Acid (OCT), Decanoic Acid (DEC), Thymol (THY), Phenylacetic acid (PhAA), Methanol, Ethanol were purchased from Merck (Darmstadt, Germany) and Alfa-Aesar (Haverhill, MA, USA) and were used without further purifications (purities > 98.5%). Hygroscopic reagents were desiccated under P_2O_5 prior use. Trimethylbenzylammonium methanesulfonate was synthesized following a procedure reported elsewhere [26]. Water was used at milliQ purity grade (>18 MΩ).

A Sartorius LE225D was used as analytical balance; the centrifugations were performed using a Beckmann Coulter ALLEGRA 64R Centrifuge; Agilent 8453 UV-VIS Spectroscopy system equipped with a thermostat (25.0 ± 0.1 °C) was used for the UV-VIS spectra determination.

HPLC-grade and MS-grade acetonitrile (ACN, purity > 99.9%) and formic acid (FA purity ≥ 95%) were purchased from Sigma Aldrich (Milan, Italy). Water for HPLC analysis was purified with a Milli-Q Plus185 system from Millipore (Milford, MA, USA). The HPLC-UV study was performed on a Thermo Separation low-pressure quaternary gradient pump system (Spectra system Series, Thermo Scientific, Waltham, MA, USA) supplied with a GT-154 vacuum degasser (Shimadzu, Kyoto, Japan). The system was equipped with a SPD-10A UV-Vis detector (Shimadzu, Kyoto, Japan) and a Rheodyne 7725i injector (Rheodyne Inc., Cotati, CA, USA) with a 20 µL stainless steel loop. Data management and acquisition was made by means of Clarity Lite chromatography software. UV detection was carried out at 360 nm. A Robusta RP18 (250 × 4.6 mm i.d., 5 µm, 100 Å pore size from Sepachrom, Milan, Italy) was used as analytical column. A Grace (Sedriano, Italy) heater/chiller (Model 7956R) thermostat was used to carry out the RP-HPLC analyses at a column temperature fixed at 25 °C. All the analyses were carried out at a 1.0 mL min^{-1} flow rate. For UHPLC-MS/MS analysis an Agilent 1290 Infinity LC system coupled with an Agilent 6540 UHD Accurate Mass QTOF (Agilent Technologies, Santa Clara, CA, USA) with an Agilent Jet Stream Dual electrospray (Dual AJS ESI) interface was used. VELP Scientifica AREX oil bath with a VTF Vertex was used for the heating and the stirring of the samples, Branson BRANSONIC 220 sonicator bath (75 W sonication power) was used for the sonication procedure. The analytes separation was performed with a Kinetex (100 × 2.1 mm i.d., 1.7 µm, 100 Å) column from Phenomenex (Torrance, CA, USA) connected with a guard cartridge EVO-C18 (2.1 × 2 mm) from Phenomenex.

2.2. DESs Preparation and Water Dilutions

The Deep Eutectic Solvents were prepared by mixing and heating (~70–80 °C) the weighted components in a sealed flask until homogeneous fluids were obtained in a timeframe spanning from 10 min to 3 h [29]. The water dilutions were prepared by adding the specific weighted amounts of water to the DESs and then leaving them under magnetic stirring at 25 °C overnight in order to generate homogenous fluids [38]. The water content of the starting mixtures was measured with a Karl Fischer titrator (Metrohm 684 KF Coulometer) and the values were found to span from 0.1 to 5% w/w in the different DESs: Ethylene Glycol/Choline Chloride (EG/ChCl, 2/1 molar ratio) 1.9% w/w; Glycerol/Choline Chloride (GLY/ChCl, 2/1 molar ratio) 3.1% w/w; Urea/Choline Chloride (U/ChCl, 2/1 molar ratio) 1.6% w/w; Glycerol/Trimethylglycine (Gly/TMG, 3/1 molar ratio) 3.6% w/w; Glycolic Acid/Trimethylglycine (GA/TMG, 2/1 molar ratio) 1.9% w/w; Glycolic Acid/L-Proline (GA/L-Pro, 3/1 molar ratio) 2.1% w/w; Glycolic Acid/Choline Chloride (GA/ChCl, 2/1 molar ratio) 2.4% w/w; p-toluenesulfonic acid/benzyltrimethylammonium methanesulfonate (pTSA/BZA, 1/1 molar ratio) 4.6% w/w; Thymol/Decanoic Acid (THY/DEC, 2/1 molar ratio) 0.5% w/w; Phenylacetic Acid/Trimethylglycine (PhAA/TMG, 2/1 molar ratio) 1.6% w/w; Thymol/Trimethylglycine (THY/TMG, 3/1 molar ratio) 0.4% w/w; Phenylacetic Acid/N,N-dimethyl-N,N-didodecylammonium chloride (PhAA/DDDACl, 2/1 molar ratio) 2.2% w/w.

2.3. Heating-Ultrasound Assisted Extraction Procedure

The onion skin leaves were weighted in a vial then the DES was weighted in the same recipient. The samples were put in an oil bath at the proper temperature under magnetic stirring at 300 rpm. The sonication procedure was made by putting the samples in the ultrasound bath at room temperature (20–25 °C) for the established (evaluated) time period. Samples were then centrifugated at 25 °C for 30 min at 7000 rpm. A total 50 µL of the orange/red supernatant (Gilson P-100 pipette) was dissolved in 2 mL of ethanol in quartz cuvette for the UV-VIS analysis in the 190 nm to 1100 nm wavelength range; spectra were normalized with Microsoft Excel software. Then, 100 µL of the same supernatant was dissolved in 2 mL of ethanol for HPLC and LC-MS/MS analysis. All the samples were analyzed in triplicate and the errors evaluated via standard deviation of the three samples.

2.4. Onion Skin Samples

Dry onion skins from the "Dorata di Parma" cultivar were bought in a local market and processed without any further pre-treatments. The onion skin samples were desiccated under vacuum using a KNF Laboport solid PTFE vacuum pump at room temperature (20–25 °C) away from sunlight, in times spanning from 1 to 6 h until constant weight. The samples were always kept in closed containers away from light and sunlight.

In order to ensure the proper comparison of the obtained results, the sets of experiments were performed with the same batches of finely chopped onion skin (about 1 mm^2).

2.5. RP-HPLC-UV Analysis

The extraction efficiencies afforded by each of the DES mixtures were evaluated through HPLC-UV-Vis analysis, by relying upon a gradient program slightly modified from a previously developed and optimized method [59]. The final gradient program was obtained from eluent A (0.1% (v/v) FA in water) and eluent B (0.1% (v/v) FA in ACN) as follows: 0 min 100% A, 0–5 min from 100% up to 97% A, 5–45 min from 97% up to 50% A, 45–50 min 0% A. At the end of each run, a column cleaning of 10 min with 100% B was added before column re-equilibration with 100% A.

2.6. UHPLC-MS/MS

The UHPLC analyses were performed under gradient conditions. Eluent A was water containing 0.1% (v/v) FA and eluent B methanol containing 0.1% (v/v) FA. The gradient was as follows: 0 min 1% B, 5 min 3% B, 45 min 50% B, 53 min linear gradient 100% B, and a post run time of 3 min to return to initial condition and re-equilibrate the system. The flow rate was 0.4 mL min^{-1}, the injection volume was 5 µL and the column temperature was 40 °C. UV-DAD spectrum range was included between 190 and 630 nm.

The acquisition was performed in positive mode. The Dual AJS ESI gas temperature was set at 350 °C. The sheath gas temperature at 400 °C, the gas flow and the sheath gas flow at 9 mL min^{-1}, the nebulizer at 35 psig, the capillary voltage at 4000 V, the nozzle voltage 0 V, the fragmentor at 120 V, the skimmer at 65 V and the Octopole RF Vpp at 750 V.

2.7. Quercetin Recovery

The recovery of the raw materials extracted was performed applying two different methods. With the anti-solvent procedure at the end of the centrifugation the material was filtered in folded paper filter and then diluted with water (or with 10% HCl—by volume—water solution in case of acidic dilutions for the hydrolysis of glycosylated quercetin) in a way to obtain 75% w/w water solution of the DES. Samples were left stirring (350 rpm) at room temperature (20–25 °C) overnight in order to permit the DES de-structuration and the HBD-HBA bond cleavage. The solutions were then centrifuged, the pellets collected and dried under vacuum with P_2O_5 and the supernatant filtered in weighted Sartorius 0.2 µm filters. The reversed-phase chromatography recover procedure was performed on the 75% w/w water (or 10% v/v HCl) solutions in Supelco Supelclean LC-18 SPE Tubes with water (3 mL) followed by methanol (3 mL).

Both the procedures were performed starting from 0.2 g of onion skin. The quercetin amounts in these crudes were then determined via HPLC analyses.

3. Results and Discussion

3.1. Optimal Deep Eutectic Solvent Design for the Quercetin Extraction

The first step of this work was the choice of the optimal solvent for the extraction of quercetin from the onion skin waste. The extraction was performed by a heating/stirring- and ultrasound-assisted procedure. In order to investigate the effectiveness of the different solvents we have firstly set-up experimental conditions commonly reported in literature [63]: heating at 50 °C and stirring at 300 rpm for 30 min then 45 min of sonication in bath followed by centrifugation of the extracts for 30 min at 7000 rpm. Afterwards, the extraction conditions were optimized as well. Because of the large number of samples to be analyzed, an UV-Vis spectra analysis of the extracts at the same dilutions was performed in turn facilitating the rapid evaluation of the extraction efficiencies of the different liquids. Absorbance values were recorded at 300 nm, which is the wavelength used to monitor the presence of molecules such as protocatechuic acid, and 366 nm, which is the typical wavelength used to study flavonoids such as quercetin and their glycosylated forms.

Differently structured DESs were used in this set of experiments, starting from commonly used ones (i.e., Urea/Choline Chloride or Glycerol/Choline Chloride mixtures and so on) [64,65] moving to other differently structured acid-based DESs (such as i.e., Glycolic Acid/Betaine, pTSA-based or Glycolic acid/L-Proline mixtures) [26,30,66].

We chose these liquids aimed at investigating the effect by neutral forms (glycerol- or glycol-based) as well as slightly acidic (Glycolic Acid-based) and highly acidic ones (p-toluensulphonic acid based). This heterogeneous selection intended to appraise whether a form was more capable than another to favor the hydrolysis of the glycosylated forms of the extracted flavonoids. Moreover, hydrophobic DESs mixtures were tested to evaluate the water solubility effect of the liquids on the extraction efficacy, considering the low solubility of the quercetin itself in water [18,67].

The liquids tested were: Ethylene Glycol/Choline Chloride (EG/ChCl, 2/1 molar ratio); Glycerol/Choline Chloride (GLY/ChCl, 2/1 molar ratio); Urea/Choline Chloride (U/ChCl, 2/1 molar ratio); Glycerol/Trimethylglycine (Gly/TMG, 3/1 molar ratio); Glycolic Acid/Trimethylglycine (GA/TMG, 2/1 molar ratio); Glycolic Acid/L-Proline (GA/L-Pro, 3/1 molar ratio); Glycolic Acid/Choline Chloride (GA/ChCl, 2/1 molar ratio); p-toluenesulfonic acid/benzyltrimethylammonium methanesulfonate (pTSA/BZA, 1/1 molar ratio); Thymol/Decanoic Acid (THY/DEC, 2/1 molar ratio); Phenylacetic Acid/Trimethylglycine (PhAA/TMG, 2/1 molar ratio); Thymol/Trimethylglycine (THY/TMG, 3/1 molar ratio); Phenylacetic Acid/N,N-dimethyl-N,N-didodecylammonium chloride (PhAA/DDDACl, 2/1 molar ratio). In addition to the above mixtures, neat methanol was also used as extraction solvent for comparative purposes. Indeed, this alcohol is commonly used for the polyphenol extraction from natural sources [68,69]. In order to evaluate the advantages of the aqueous solutions of DESs (such as the fine tuning of the overall viscosity and the polarity extent), water additions were also tested. DESs undergo structural changes by adding water in amounts that are dependent on the structural features of the components and on their interactions. Thus, three different dilutions were scrutinized using three different amounts of water (10%, 30% and 70% w/w) in each solvent; considering that with starting water amounts from 0.1% to 5% in the liquids, with these values it is possible to easily stay between the values of water that determine structural changes in the DESs solutions [38,70–72]. The hydrophobic DESs were tested as such, without any water addition: indeed, they generally do not absorb water contents higher than 10% w/w [18]. Water dilutions of methanol were also tested. In Figure 1, the results of absorbance of the samples at 366 nm are reported, while the absorbance values at 300 nm of the same samples are reported in the Supplementary Materials (Figure S1) as well as UV-Vis spectra of a typical sample and of all the samples (Figure S2).

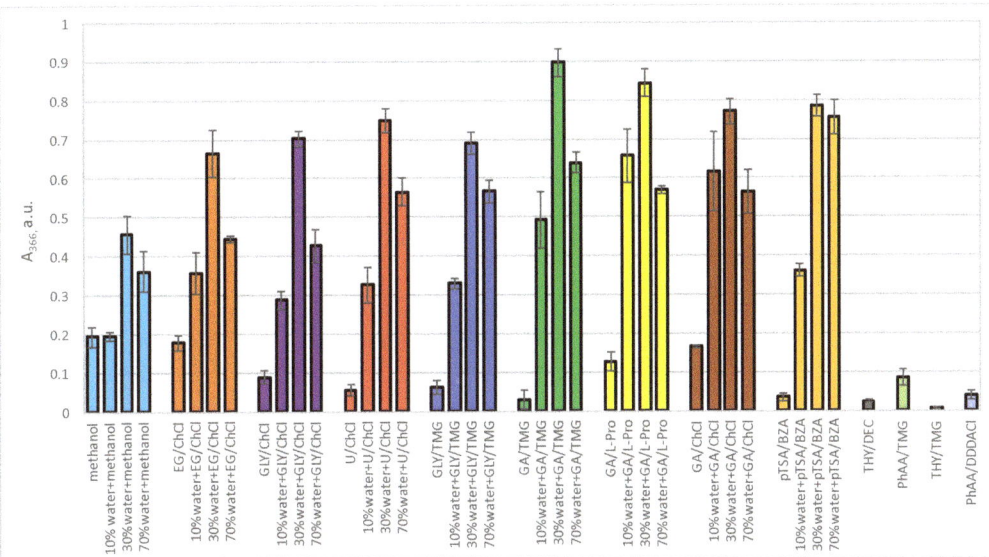

Figure 1. UV-Vis Absorbance at λ = 366 nm of the supernatants of the extraction of onion skin diluted in ethanol (50 µL in 2 mL EtOH). Extraction conditions: 50 mg of onion skin in 1.5 g of aqueous DES, heating and stirring (50 °C, 300 rpm) for 30 min then 45 min of sonication in bath followed by centrifugation of the extracts for 30 min at 7000 rpm. EG/ChCl 2/1 molar ratio; GLY/ChCl 2/1 molar ratio; U/ChCl 2/1 molar ratio; Gly/TMG 3/1 molar ratio; GA/TMG 2/1 molar ratio; GA/L-Pro 3/1 molar ratio; GA/ChCl 2/1 molar ratio; pTSA/BZA 1/1 molar ratio; THY/DEC 2/1 molar ratio; PhAA/TMG 2/1 molar ratio; THY/TMG 3/1 molar ratio; PhAA/DDDACl 2/1 molar ratio. Water amounts are considered as added water to the starting DESs (initial water amounts spanning from 0.1% to 5% w/w).

From the UV-Vis analyses on the raw extracted material, it is evident that all the pure non-diluted DESs have extraction efficacies lower than methanol. EG/ChCl DES showed the highest extraction efficacy in its pure non-diluted form compared to the other pure liquids. However, these values increase steeply by addition of water: the absorbance values were more than doubled with 10% w/w water, with the highest values recorded at 30% w/w; then, a decrease did occur with 70% w/w of added water. This trend is coherent with the structural properties of the water dilutions of DESs as the lowering of the liquid viscosity, which follows the increasing amounts of water promoting an easier mass transfer (that is, the extraction power) from the onion skin. This is also suggested by the low solubility of quercetin (or similar phenols contained in the onion skin) in water, that therefore could be easily extracted from onion skin thanks to the lower viscosity and then it could be solubilized in the DESs domains. The HBD-HBA bond cleavage and following DESs structures disruptions at values of 70% w/w of added water led to a decrease on the extraction efficiency, even though it remained higher than that produced both by neat methanol and water-methanol solutions. The negative inflection of the extraction trend could be due to interactions of the phenols with the DESs isolated components. The changes in the polarity of the media did not play a significant role as the trend of the A_{366} of the water dilutions was the same for all the differently structured polarities. The water additions, in fact, lead to changes in polarity that lead progressively to the polarity of water itself by increasing its amount [38]. In this case some of the solvents could have higher polarity than water and other lower ones, so the trends could have been different. The hydrophobic DESs (THY/DEC, PhAA/TMG, THY/TMG, PhAA/DDDACl) showed lower extraction efficacies as they could not benefit from the advantaged resulting from the water addition.

The acidity of the HBD in the DESs plays a role in their extraction efficiency as the common non-acid liquids (EG/ChCl, GLY/ChCl, U/ChCl, Gly/TMG) showed absorbances at 366 nm lower than the ones with acidic HBDs such as glycolic acid-based liquids (GA/TMG, GA/L-Pro, GA/ChCl) and the pTSA-based one (pTSA/BZA). HBA seems to play a less relevant role as the most important differences were observed by changing the HBD.

The best extracting liquid in our set was found to be the glycolic acid/betaine (GA/TMG) mixture, already known in the literature for its multiple advantageous uses [73].

The absorbances at 300 nm were higher than those at 366 nm, but the A_{366}/A_{300} trend was identical for all the samples, therefore suggesting the non-selective extraction of the different DESs in the set.

From the data reported in Figure 1 it emerges that all the DESs in their water dilutions at 30% (w/w) are much more efficient than methanol. In particular GA/TMG showed more than 4.5 times higher extraction efficiency of the raw material compared with methanol and over twice compared with methanol with 30% (w/w) water. These data strongly promote the use of the water dilutions of DESs as an efficacious and green method for the extraction of important phenolic compounds from dry onion skin. In Figure S3 in Supplementary Materials, the ratio of A_{366} of the samples on the A_{366} of pure methanol and of methanol with 30% (w/w) of added water are reported in order to evaluate the efficacy of the DESs water dilutions compared to methanol (the same data at 300 nm is reported in the same figure).

3.2. Extraction Conditions Optimization

Once the optimal DESs were found, the optimization of the extraction conditions was made by changing (i) the amount of water added to the DESs extracting liquids (this time, a more extended water content was evaluated), (ii) the time of heating, (iii) the temperature of heating and (iv) the sonication time. All these procedures were performed sequentially, according to the one-variable-at-time (OVAT) approach. These experiments were performed with the system GA/TMG as this DES emerged as best performing one for the scope of the present work. Moreover, it is also characterized by a low cost and easy preparation. In these experiments, the estimation of the extraction efficacy from the dry onion skin was also evaluated via UV-Vis analysis at 366 nm. The UV-Vis absorbances at 300 nm gave the same trends and therefore these values are not reported herein. In Figure 2, the results of the optimization steps are reported; the optimization of the ratio of the mass of onion skin on the mass of the extracting DESs was also performed, but it did not show any relevant trend (see Supplementary Materials, Figure S4). Therefore, the following amounts were maintained also in this part of the study: 50 mg of onion skin with 1.5 g of DES.

The first parameter considered was the amount of water added to the DESs (Figure 3A); in this framework, a set of ten experiments was performed from 0% to 90% (w/w) added water. The best result obtained in terms of absorbance at 366 nm was at 30% w/w, the same value that turned out in the previous step dealing with the screening of the various DESs. With this water amount the heating time at 50 °C was then evaluated in a time-frame of 120 min (Figure 3B). The best result was again that previously identified during the screening stage, that is, 30 min. Then, the heating temperature was varied in the range 25 °C–100 °C (Figure 3C). In this case, an almost constant value of A_{366} was observed from 50 °C to 80 °C, then an increase was recorded corresponding to a browning of the extracting solution and difficult operating conditions (impossible separation of the onion matrix after centrifugation). Therefore, the temperature was set as optimal at 50 °C because it is the lowest temperature that gave the optimal results without browning of the solutions. The only parameter that was changed from the starting conditions was the sonication time because it showed an increase of A_{366} without any experimental drawback (Figure 3D): an increase from 45 min to 1 h showed an increase of the extracted raw material. Therefore, the optimal extraction conditions were: 50 mg of onion skin in 1.5 g of 30% w/w of added water in DESs, heating and stirring at 50 °C for 30 min, followed by sonication of 1 h and then centrifugation of the sample for 30 min at 7000 rpm.

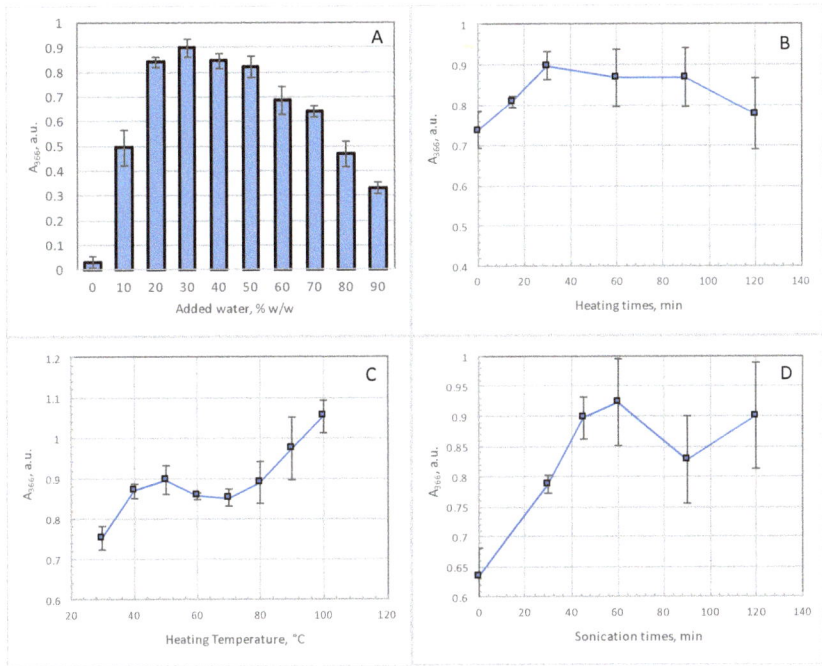

Figure 2. Optimization of the extraction procedure, GA/TMG DES + 30% w/w added water, 50 mg of onion skin in 1.5 g of DES. (**A**): optimization of the water amount on GA/TMG DES, heating and stirring (50 °C, 300 rpm) for 30 min then 45 min of sonication in bath followed by centrifugation of the extracts for 30 min at 7000 rpm. (**B**): optimization of heating times, heating and stirring (50 °C, 300 rpm) then 45 min of sonication in bath followed by centrifugation of the extracts for 30 min at 7000 rpm. (**C**): optimization of heating temperature, heating and stirring (300 rpm) for 30 min then 45 min of sonication in bath followed by centrifugation of the extracts for 30 min at 7000 rpm. (**D**): optimization of sonication times, heating and stirring (50 °C, 300 rpm) for 30 min then sonication in bath followed by centrifugation of the extracts for 30 min at 7000 rpm. All the measures are averages of triplicates and the error bars are standard deviations of the set.

The samples of GA/TMG, GA/L-Pro and pTSA/BZA, obtained by applying these experimental conditions, were then submitted to HPLC analysis (see Section 3.3 for details). Methanol and its water dilutions were also considered for comparative purposes.

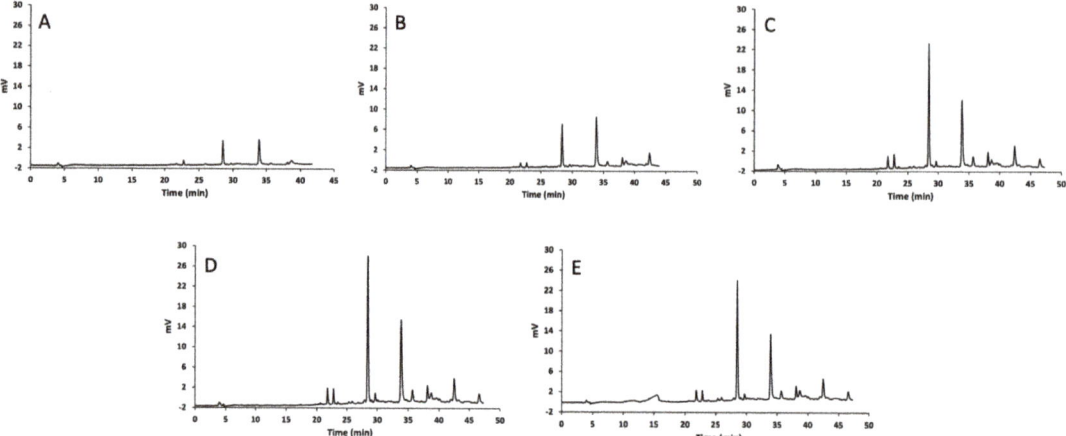

Figure 3. Chromatograms of the HPLC-UV-Vis analysis on onion extracts at 360 nm. (**A**): pure MeOH, (**B**): MeOH 30% w/w water, (**C**): GA/TMG +30% w/w water, (**D**): GA/L-Pro + 30% w/w water, (**E**): pTSA/BZA + 30% w/w water.

3.3. RP-HPLC-UV Analysis

In order to characterize the qualitative and quantitative profile of each onion extract, a HPLC-UV (wavelength of detection 360 nm) analysis was firstly carried out followed by an UHPLC-MS investigation for the identification of the main peaks.

The chromatographic profiles clearly evidenced the presence of two main peaks (Figure 3). The first one, with a retention time of about 29 min, was plausibly ascribed to a glycosylated form of quercetin and successively confirmed and characterized by UHPLC-MS analysis (see Section 3.3.1 for details). In fact, besides the aglycone of quercetin, diglucosides (mainly quercetin-3,4'-O-diglucoside) and glucosides derivatives (mainly quercetin-4'-O-diglucoside) represent the predominant forms in different onion varieties [74–76]. Moreover, the prevalence of the mono-glycosilated quercetin with respect to the di-glycosylated form could reasonably originate from a hydrolytic cleavage occurring during the extraction. The second main peak was identified as the quercetin aglycone, based on the correspondence of peak retention time (around 34 min) with that of the reference standard.

The applied gradient program produced a profitable separation of the selected peaks from other minor compounds or matrix interferences. This in turn allowed the reliable quantitation of quercetin, its extraction being the focus of the present study. A noteworthy major content of quercetin was always recovered with the use of different DESs mixtures if compared to more conventional extraction protocols operated with pure methanol or its hydro-alcoholic mixtures.

The exemplary chromatograms of onion skin extracts submitted to conventional methods or DES-based extraction protocols are shown in Figure 3. The results also evidenced a higher content of glycosylated quercetin provided by DESs extractants, thus underlying their effectiveness and selectivity towards such class of flavonoids with respect to traditional methods.

3.3.1. UHPLC-MS/MS

According to HPLC results, the UHPLC-MS/MS analysis was focused on the identification of peak at lower retention time and with UV absorption at 366 nm. The sample analyzed was the one extracted with GA/L-Pro + 30% added water liquid. The results are reported in Figure 4.

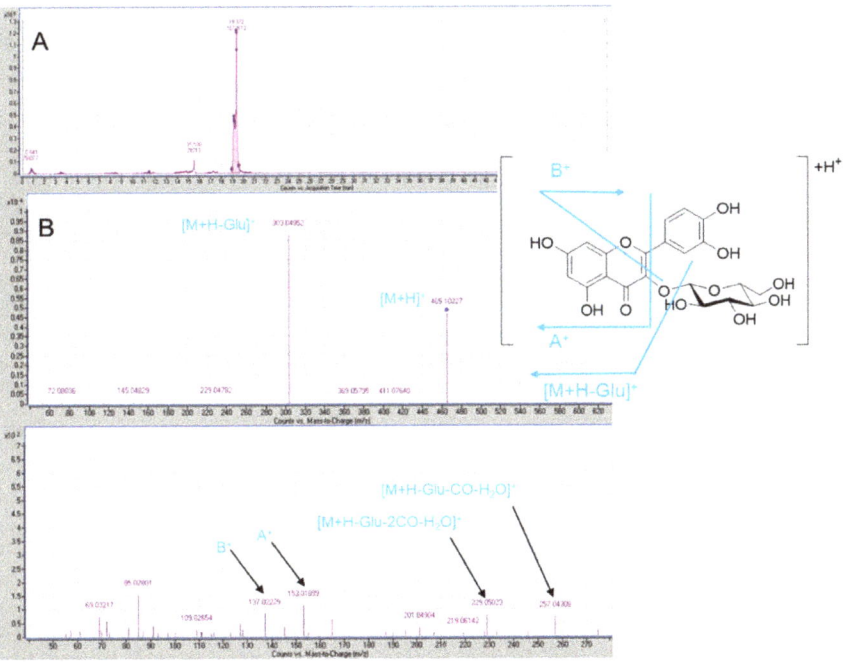

Figure 4. LC-MS/MS analysis of the peak at lower retention time of GLY/L-Pro + 30% added water sample. (**A**): LC chromatogram; (**B**): MS and MS/MS fragmentations of the peak at 19 min retention time.

The MS spectra show a peak with m/z value of 465.1029 that corresponds to pseudomolecular ion $[M+H]^+$ of a compound with chemical formula $C_{21}H_{20}O_{12}$. The $[M+H]^+$ fragmentation pattern is consistent with the loss of a glycosyl or galactosyl due to the presence of the fragment with m/z 303.0493. The presence of fragments with m/z at 257.0430, 229.0502, 153.0189, 137.0223 are related to the fragmentation of quercetin moiety as reported in literature [77,78] and they confirm that the compound is a glycosylated form of quercetin.

3.3.2. Quantitation of Quercetin in the Investigated Extracts

The quantitation of quercetin in all the investigated extracts was performed by relying upon a calibration curve built up by using standard solutions with concentration values spanning in the range specified in Table S1 (Supplementary Materials). As evident by the R^2 value, the obtained mathematical models were characterized by a very good linearity. The established HPLC method was further validated in terms of accuracy, precision, and limit of detection (LOD) and limit of quantification (LOQ) (Tables S1 and S2, Supplementary Materials). Accordingly, high recovery% values (from 98.15% up to 99.56%) and low range of variation of the RSD% values (from 0.83% up to 1.04%) were observed when the long-term (inter-day) accuracy and precision were evaluated, respectively (Table S2, Supplementary Materials). Additionally, appreciably low LOD (0.37 µg/mL) and LOQ (1.11 µg/mL) values were calculated for quercetin samples (Table S1, Supplementary Materials). The obtained results are comparable with data reported in literature [79,80]. For example, the validation data of the HPLC method for quercetin determination in green tea reported by Savic and co-workers showed accuracy values between 98.2 and 101.3%, RSD% values in the range 0.89–1.55%, and LOD and LOQ values of 1.2 and 4.0 µg/mL, respectively [81].

The consistent and reliable outcomes achieved with the validation process revealed the adequacy of the analytical method to be applied for quantitative purposes.

The results shown in Table 1 highlight a certain selectivity in terms of extracted compounds by DESs characterized by a higher content of quercetin in all the extracts with respect to the traditional media. In particular, the mixture GA/L-Pro produced the highest quercetin recovery. Interestingly, it is worth noting that the ratio between the glycosylated form and the aglycone one was kept quite constant when extractions were performed with DESs systems. In these cases, the ratio between the two forms was around 1.4 with a slight prevalence of the glycosylated form over the aglycone one. On the contrary, despite the inverted ratio found when pure methanol or a methanol/water mixture was employed, a lower total recovery was reached.

Table 1. Quantitation of quercetin amounts and glycosylated/aglycone quercetin ratio in the analyzed samples.

Onion Extract	Quercetin Mean Conc. ± SD (µg/mL)	Glycosylated/Aglycone—Quercetin Ratio
MeOH	5.84 ± 0.13	43/57
MeOH + 30% w/w water	10.83 ± 0.01	40/60
GA/TMG + 30% w/w water	14.79 ± 0.50	58/42
GA/L-Pro + 30% w/w water	18.56 ± 0.25	58/42
pTSA/BZA + 30% w/w water	14.83 ± 0.31	59/41

3.4. Quercetin Recovery

The recovery of the quercetin from the extraction matrix was performed with two different methodologies: water anti-solvent method and SPE (solid phase extraction) method. In both protocols it was considered that at water amounts over 75% (w/w) the DESs structures are disrupted and the bonds HBA-HBD are cleaved; in these conditions the limited water solubility of quercetin can be exploited for its separation from the hydrophilic media. Therefore, the experiments were carried out in the optimized conditions and then water was added in order to get to 75% (w/w) of added water. Then, the samples were left stirring at room temperature (20–25 °C) overnight and finally treated in the two different methodologies. With anti-solvent technique, the samples were centrifuged and the solid phases separated, the supernatants filtered in weighted 0.2 µm filters and the solid phases diluted in the proper amounts of ethanol to perform the HPLC quantitative analyses. The SPE was performed via loading of the sample in the reversed-phase cartridge and then the products were recovered with methanol (3 mL) wash. The same experiments were conducted with HCl 10% (w/w) water solution instead of water; in this way an increase of the non-glycosylated form could be collected because of acid hydrolysis. The anti-solvent procedure gave 0.16 g of raw material starting from 0.2 g of onion leaves and the reverse-phase chromatography gave 0.016 g of raw material starting from 0.2 g of onion skin.

The yields of quercetin recovery, calculated as amount of quercetin obtained on the amount of quercetin extracted both evaluated via HPLC analyses, are reported in Table 2. In the same table, the data coming from the experiments conducted with HCl 10% (w/w) are reported in terms of ratio glycosylated/non-glycosylated forms. This is because in these cases it is not possible to calculate a yield because of the increasing amount of non-glycosylated quercetin.

The SPE (solid phase extraction) method gave excellent yield of recovery of quercetin (81%), while the water anti-solvent method showed low performances of recovery (8%). The low values observed with anti-solvent method could be due to interactions occurring between the DES components (glycolic acid and trimethylglycine) and the quercetin that could lead to more water-soluble adduct not allowing the precipitate formation after centrifugation. Polyphenols extraction procedures made by DESs are known to occur thanks to weak interactions occurring between the phenols and the network of weak forces in the DESs liquids [24,82]. This is supported by the values of A_{366} in extraction procedures observed with 70% (w/w) of added water in the optimal DES design that were in fact still over the methanol extraction even if lower than the maximum observed at 30% (w/w) added water.

Table 2. Recovery of quercetin from the extracted samples evaluation via HPLC analyses.

Recovery Procedure	Sample	Quercetin Mean Conc. ± SD (µg/mL)	Yield of Recovery, %	Glycosylated/Aglycone—Quercetin Ratio H_2O Recover	Glycosylated/Aglycone—Quercetin Ratio HCl 10% w/w Recover
SPE	extracted recovered	12.83 ± 0.01 [a] 11.68 ± 0.39 [b]	81%	52/48	40/60
Anti-Solvent	extracted recovered	13.88 ± 0.02 [c] 4.49 ± 0.01 [d]	8%	38/62	29/71

Extracting liquid GA/TMG + 30% w/w added water (density 1.1941 g/mL), heating and stirring at 50 °C for 30 min, 1-h sonication, centrifugation of the sample for 30 min at 7000 rpm, filtration of the sample with water amounts to give 75% w/w of added water left stirring overnight. Yields of recovery calculated as percent of recover from the extracted material reported at the same dilutions. Dilutions made in order to give values of areas of HPLC analyses inside the calibration curve. Glycosylated/Aglycone—Quercetin Ratio calculated as ratio of HPLC peaks area in the recover procedure with water (H_2O recover) or with 10% w/w HCl in water solution (HCl 10% w/w recover column). [a] = 50 µL of sample from 1.2587 mL extracting DES batch dissolved in 2 mL of EtOH; [b] = 140 µL of sample from a total 2 mL EtOH batch dissolved in 2 mL of EtOH; [c] = 50 µL of sample from 5.3641 mL extracting DES batch dissolved in 2 mL of EtOH; [d] = 200 µL of sample from a total 5 mL EtOH batch dissolved in 2 mL of EtOH.

In the SPE method, excellent values of recovery were observed (81%) but methanol was used for the recovery after the wash in the SPE cartridge. However, the amounts of extracted material with the DESs water dilutions showed values that are over 4.5 and 2 times higher than the ones of methanol or methanol with 30% (w/w) of added water when used as extracting agents. If the recovery efficacy is normalized on the amounts of methanol used (1.89 mL in case of extraction with methanol and 3 mL in case of DES) the procedure is still advantageous because the extraction efficacy is 4.5 times higher with almost twice the methanol used, therefore it is almost three times more efficacious. Moreover, when used as extracting liquid, methanol is heated to temperatures close to its boiling point, therefore implying peculiar attention to the experimental conditions (aspirating hoods, flammability of the media, toxicity of the vapors and so on) that increase in the case of industrial scale-up of the process. This feature undoubtedly promotes the DES-water system for the inherent extraction efficiency and overall greenness of the method.

The use of HCl 10% (w/w) water solutions instead of the simple water for the dilutions, led to an increase of the amount of non-glycosylated form as expected; in this case, the yields of recovery was not calculated as the amounts of quercetin recovered were also higher than the ones initially extracted.

4. Conclusions

In this work water dilutions of a set of Deep Eutectic Solvents (DESs) revealed to be excellent and efficacious green media for the selective extraction of quercetin and its glycosylated form from onion skin, a low-cost waste material. As reported, the best results were obtained with the use of acidic components in the DESs liquids (GA/TMG, GA/L-Pro and pTSA/BZA). Glycolic acid-based ones can be considered NADESs, therefore their greenness is increased over the pTSA-based one that moreover has strong acidity in its components. However, no effect on the ratio aglycone/glycosylated forms was observed by changing the acidic strength in the liquids even though the O-glycosidic bond can usually be hydrolyzed in acidic conditions, therefore suggesting a different mechanism of extraction. The procedure revealed to be much more effective than the use of methanol, a highly toxic and volatile solvent commonly used in the extraction of polyphenols from vegetal matrixes. The quercetin concentration in the samples (in the aglycone form only) were in fact over three times higher than methanol as emerged from HPLC analyses (5.84 µg/mL with methanol compared to 18.56 µg/mL with GA/L-Pro and over 14 µg/mL for GA/TMG and pTSA/BZA samples) and more than 1.5 times higher using the water/methanol mixture (10.83 µg/mL). The extracted materials were also recovered efficaciously with solid phase extraction method with excellent yields (81%) of recovery.

The proposed extraction protocol revealed to be green, efficacious and selective for the extraction from onion skin of quercetin, a molecule that is gaining importance for properties such as its pharmacological activity.

Supplementary Materials: The following are available online at https://www.mdpi.com/article/10.3390/ma14216465/s1, Figure S1: UV-Vis Absorbance at λ = 300 nm of the supernatants of the extraction of onion skin. Figure S2: UV-Vis spectra of MeOH, MeOH +30% added water and GA/TMG + 30% added water samples—UV-Vis spectra of all the extraction samples. Figure S3: ratio of UV-Vis A_{300} nm and A_{366} nm sample/methanol and sample/methanol + 30% w/w water added. Figure S4: ratio of the mass of onion skin on the mass of the extracting DESs optimization. Table S1: HPLC Calibration data. Table S2: HPLC Method validation.

Author Contributions: Conceptualization, M.T., C.C., R.G. and R.S.; methodology, M.T. and R.G.; software, M.T.; validation, R.G., C.C. and M.T.; formal analysis, M.T. and C.C.; investigation, M.C., F.I. and S.D.B.; resources, C.C. and R.G.; data curation, M.T. and M.C.; writing—original draft preparation, M.T., F.I. and S.D.B.; writing—review and editing, R.S., R.G., C.C. and L.C.; visualization, M.T. and R.S.; supervision, M.T., R.G. and C.C.; project administration, M.T. and R.G.; funding acquisition, C.C. and L.C. All authors have read and agreed to the published version of the manuscript.

Funding: This research was funded by Fondazione Cassa di Risparmio di Perugia, Grant number 220.0513.

Institutional Review Board Statement: Not applicable.

Informed Consent Statement: Not applicable.

Data Availability Statement: The data presented in this study are available on request from the corresponding author.

Acknowledgments: The authors thank Filippo Bocerani for the technical support.

Conflicts of Interest: The authors declare no conflict of interest.

References

1. DeSimone, J.M. Practical Approaches to Green Solvents. *Science* **2002**, *297*, 799–803. [CrossRef] [PubMed]
2. Moran, M.J.; Zogorski, J.S.; Squillace, P.J. Chlorinated Solvents in Groundwater of the United States. *Environ. Sci. Technol.* **2007**, *41*, 74–81. [CrossRef]
3. Ikeda, M. Solvents in urine as exposure markers. *Toxicol. Lett.* **1999**, *108*, 99–106. [CrossRef]
4. Pearson, J.K. European solvent VOC emission inventories based on industry-wide information. *Atmos. Env.* **2019**, *204*, 118–124. [CrossRef]
5. Paiva, A.; Craveiro, R.; Aroso, I.; Martins, M.; Reis, R.L.; Duarte, A.R.C. Natural Deep Eutectic Solvents–Solvents for the 21st Century. *ACS Sustain. Chem. Eng.* **2014**, *2*, 1063–1071. [CrossRef]
6. Florindo, C.; Lima, F.; Ribeiro, B.D.; Marrucho, I.M. Deep eutectic solvents: Overcoming 21st century challenges. *Curr. Opin. Green Sustain. Chem.* **2018**, *18*, 31–36. [CrossRef]
7. Alonso, D.A.; Baeza, A.; Chinchilla, R.; Guillena, G.; Pastor, I.M.; Ramón, D.J. Deep Eutectic Solvents: The Organic Reaction Medium of the Century. *Eur. J. Org. Chem.* **2016**, *2016*, 612–632. [CrossRef]
8. Alonso, D.A.; Burlingham, S.; Chinchilla, R.; Guillena, G.; Ramón, D.J.; Tiecco, M. Asymmetric Organocatalysis in Deep Eutectic Solvents. *Eur. J. Org. Chem.* **2021**, *2021*, 4065–4071. [CrossRef]
9. CAraujo, C.F.; Coutinho, J.A.P.; Nolasco, M.M.; Parker, S.F.; Ribeiro-Claro, P.J.A.; Rudić, S.; Soares, B.I.G.; Vaz, P.D. Inelastic neutron scattering study of reline: Shedding light on the hydrogen bonding network of deep eutectic solvents. *Phys. Chem. Chem. Phys.* **2017**, *19*, 17998–18009. [CrossRef]
10. Kaur, S.; Kumari, M.; Kashyap, H.K. Microstructure of Deep Eutectic Solvents: Current Understanding and Challenges. *J. Phys. Chem. B* **2020**, *124*, 10601–10616. [CrossRef]
11. Martins, M.; Pinho, S.P.; Coutinho, J.A.P. Insights into the Nature of Eutectic and Deep Eutectic Mixtures. *J. Solut. Chem.* **2019**, *48*, 962–982. [CrossRef]
12. Abranches, D.O.; Silva, L.P.; Martins, M.A.R.; Pinho, S.P.; Coutinho, J.A.P. Understanding the Formation of Deep Eutectic Solvents: Betaine as a Universal Hydrogen Bond Acceptor. *ChemSusChem* **2020**, *13*, 4916–4921. [CrossRef] [PubMed]
13. LKollau, L.J.B.M.; Vis, M.; Bruinhorst, A.V.D.; Esteves, A.C.C.; Tuinier, R. Quantification of the liquid window of deep eutectic solvents. *Chem. Commun.* **2018**, *54*, 13351–13354. [CrossRef]
14. Crespo, E.A.; Silva, L.P.; Martins, M.A.R.; Fernandez, L.; Ortega, J.; Ferreira, M.O.A.S.; Sadowski, G.; Held, C.; Pinho, S.P.; Coutinho, J.A.P. Characterization and Modeling of the Liquid Phase of Deep Eutectic Solvents Based on Fatty Acids/Alcohols and Choline Chloride. *Ind. Eng. Chem. Res.* **2017**, *56*, 12192–12202. [CrossRef]

15. Dutta, S.; Yu, I.K.; Tsang, D.C.; Ng, Y.H.; Ok, Y.S.; Sherwood, J.; Clark, J.H. Green synthesis of gamma-valerolactone (GVL) through hydrogenation of biomass-derived levulinic acid using non-noble metal catalysts: A critical review. *Chem. Eng. J.* **2019**, *372*, 992–1006. [CrossRef]
16. Singh, S.; Savoy, A.W. Ionic liquids synthesis and applications: An overview. *J. Mol. Liq.* **2020**, *297*, 112038. [CrossRef]
17. Cardellini, F.; Brinchi, L.; Germani, R.; Tiecco, M. Convenient Esterification of Carboxylic Acids by S N 2 Reaction Promoted by a Protic Ionic-Liquid System Formed in Situ in Solvent-Free Conditions. *Synth. Commun.* **2014**, *44*, 3248–3256. [CrossRef]
18. Tiecco, M.; Cappellini, F.; Nicoletti, F.; Del Giacco, T.; Germani, R.; Di Profio, P. Role of the hydrogen bond donor component for a proper development of novel hydrophobic deep eutectic solvents. *J. Mol. Liq.* **2019**, *281*, 423–430. [CrossRef]
19. Abranches, D.O.; Martins, M.A.R.; Silva, L.P.; Schaeffer, N.; Pinho, S.P.; Coutinho, J.A.P. Phenolic hydrogen bond donors in the formation of non-ionic deep eutectic solvents: The quest for type V DES. *Chem. Commun.* **2019**, *55*, 10253–10256. [CrossRef] [PubMed]
20. van Osch, D.J.; Zubeir, L.F.; Bruinhorst, A.V.D.; Rocha, M.A.; Kroon, M.C. Hydrophobic deep eutectic solvents as water-immiscible extractants. *Green Chem.* **2015**, *17*, 4518–4521. [CrossRef]
21. Kudłak, B.; Owczarek, K.; Namieśnik, J. Selected issues related to the toxicity of ionic liquids and deep eutectic solvents—A review. *Env. Sci. Pollut. Res.* **2015**, *22*, 11975–11992. [CrossRef] [PubMed]
22. Hayyan, M.; Hashim, M.A.; Hayyan, A.; Al-Saadi, M.A.; AlNashef, I.M.; Mirghani, M.E.S.; Saheed, O.K. Are deep eutectic solvents benign or toxic? *Chemosphere* **2013**, *90*, 2193–2195. [CrossRef]
23. Wen, Q.; Chen, J.-X.; Tang, Y.-L.; Wang, J.; Yang, Z. Assessing the toxicity and biodegradability of deep eutectic solvents. *Chemosphere* **2015**, *132*, 63–69. [CrossRef]
24. Dai, Y.; Witkamp, G.-J.; Verpoorte, R.; Choi, Y.H. Natural Deep Eutectic Solvents as a New Extraction Media for Phenolic Metabolites in *Carthamus tinctorius* L. *Anal. Chem.* **2013**, *85*, 6272–6278. [CrossRef] [PubMed]
25. Hansen, B.B.; Spittle, S.; Chen, B.; Poe, D.; Zhang, Y.; Klein, J.M.; Horton, A.; Adhikari, L.; Zelovich, T.; Doherty, B.W.; et al. Deep Eutectic Solvents: A Review of Fundamentals and Applications. *Chem. Rev.* **2020**, *121*, 1232–1285. [CrossRef]
26. De Santi, V.; Cardellini, F.; Brinchi, L.; Germani, R. Novel Brønsted acidic deep eutectic solvent as reaction media for esterification of carboxylic acid with alcohols. *Tetrahedron Lett.* **2012**, *53*, 5151–5155. [CrossRef]
27. Qin, H.; Hu, X.; Wang, J.; Cheng, H.; Chen, L.; Qi, Z. Overview of acidic deep eutectic solvents on synthesis, properties and applications. *Green Energy Env.* **2020**, *5*, 8–21. [CrossRef]
28. Di Crescenzo, A.; Tiecco, M.; Zappacosta, R.; Boncompagni, S.; Di Profio, P.; Ettorre, V.; Fontana, A.; Germani, R.; Siani, G. Novel zwitterionic Natural Deep Eutectic Solvents as environmentally friendly media for spontaneous self-assembly of gold nanoparticles. *J. Mol. Liq.* **2018**, *268*, 371–375. [CrossRef]
29. Palomba, T.; Ciancaleoni, G.; Del Giacco, T.; Germani, R.; Ianni, F.; Tiecco, M. Deep Eutectic Solvents formed by chiral components as chiral reaction media and studies of their structural properties. *J. Mol. Liq.* **2018**, *262*, 285–294. [CrossRef]
30. Tiecco, M.; Alonso, D.A.; Ñíguez, D.R.; Ciancaleoni, G.; Guillena, G.; Ramón, D.J.; Bonillo, A.A.; Germani, R. Assessment of the organocatalytic activity of chiral L-Proline-based Deep Eutectic Solvents based on their structural features. *J. Mol. Liq.* **2020**, *313*, 113573. [CrossRef]
31. Mąka, H.; Spychaj, T.; Adamus, J. Lewis acid type deep eutectic solvents as catalysts for epoxy resin crosslinking. *RSC Adv.* **2015**, *5*, 82813–82821. [CrossRef]
32. Giofrè, S.V.; Tiecco, M.; Ferlazzo, A.; Romeo, R.; Ciancaleoni, G.; Germani, R.; Iannazzo, D. Base-Free Copper-Catalyzed Azide-Alkyne Click Cycloadditions (CuAAc) in Natural Deep Eutectic Solvents as Green and Catalytic Reaction Media**. *Eur. J. Org. Chem.* **2021**, *2021*, 4777–4789. [CrossRef]
33. Alshammari, O.A.O.; Almulgabsagher, G.A.A.; Ryder, K.S.; Abbott, A.P. Effect of solute polarity on extraction efficiency using deep eutectic solvents. *Green Chem.* **2021**, *23*, 5097–5105. [CrossRef]
34. Shukla, S.K.; Nikjoo, D.; Mikkola, J.-P. Is basicity the sole criterion for attaining high carbon dioxide capture in deep-eutectic solvents? *Phys. Chem. Chem. Phys.* **2020**, *22*, 966–970. [CrossRef] [PubMed]
35. Nejrotti, S.; Iannicelli, M.; Jamil, S.S.; Arnodo, D.; Blangetti, M.; Prandi, C. Natural deep eutectic solvents as an efficient and reusable active system for the Nazarov cyclization. *Green Chem.* **2020**, *22*, 110–117. [CrossRef]
36. Dai, Y.; Witkamp, G.-J.; Verpoorte, R.; Choi, Y.H. Tailoring properties of natural deep eutectic solvents with water to facilitate their applications. *Food Chem.* **2015**, *187*, 14–19. [CrossRef] [PubMed]
37. Brinchi, L.; Germani, R.; Braccalenti, E.; Spreti, N.; Tiecco, M.; Savelli, G. Accelerated decarboxylation of 6-nitrobenzisoxazole-3-carboxylate in imidazolium-based ionic liquids and surfactant ionic liquids. *J. Colloid Interface Sci.* **2010**, *348*, 137–145. [CrossRef]
38. Gabriele, F.; Chiarini, M.; Germani, R.; Tiecco, M.; Spreti, N. Effect of water addition on choline chloride/glycol deep eutectic solvents: Characterization of their structural and physicochemical properties. *J. Mol. Liq.* **2019**, *291*, 111301. [CrossRef]
39. Chen, J.; Liu, M.; Wang, Q.; Du, H.; Zhang, L. Deep Eutectic Solvent-Based Microwave-Assisted Method for Extraction of Hydrophilic and Hydrophobic Components from Radix Salviae miltiorrhizae. *Molecules* **2016**, *21*, 1383. [CrossRef]
40. Cunha, S.C.; Fernandes, J.O. Extraction techniques with deep eutectic solvents. *TrAC Trends Anal. Chem.* **2018**, *105*, 225–239. [CrossRef]
41. Zhang, C.-W.; Xia, S.-Q.; Ma, P.-S. Facile pretreatment of lignocellulosic biomass using deep eutectic solvents. *Bioresour. Technol.* **2016**, *219*, 1–5. [CrossRef]

42. Majová, V.; Horanová, S.; Škulcová, A.; Šima, J.; Jablonsky, M. Deep eutectic solvent delignification: Impact of initial lignin. *BioResources* **2017**, *12*, 7301–7310. [CrossRef]
43. Kohli, K.; Katuwal, S.; Biswas, A.; Sharma, B.K. Effective delignification of lignocellulosic biomass by microwave assisted deep eutectic solvents. *Bioresour. Technol.* **2020**, *303*, 122897. [CrossRef] [PubMed]
44. Alañón, M.E.; Ivanović, M.; Pimentel-Mora, S.; Borrás-Linares, I.; Arráez-Román, D.; Segura-Carretero, A. A novel sustainable approach for the extraction of value-added compounds from *Hibiscus sabdariffa* L. calyces by natural deep eutectic solvents. *Food Res. Int.* **2020**, *137*, 109646. [CrossRef]
45. Zainal-Abidin, M.H.; Hayyan, M.; Hayyan, A.; Jayakumar, N.S. New horizons in the extraction of bioactive compounds using deep eutectic solvents: A review. *Anal. Chim. Acta* **2017**, *979*, 1–23. [CrossRef] [PubMed]
46. Clifford, M.N.; Madala, N.E. Surrogate Standards: A Cost-Effective Strategy for Identification of Phytochemicals. *J. Agric. Food Chem.* **2017**, *65*, 3589–3590. [CrossRef]
47. Parasuraman, S.; David, A.V.A.; Arulmoli, R. Overviews of biological importance of quercetin: A bioactive flavonoid. *Pharmacogn. Rev.* **2016**, *10*, 84–89. [CrossRef] [PubMed]
48. Karavelioğlu, B.; Hoca, M. Potential effects of onion (*Allium cepa* L.) and its phytomolecules on non-communicable chronic diseases: A review. *J. Hortic. Sci. Biotechnol.* **2021**, 1–10. [CrossRef]
49. Di Pierro, F.; Iqtadar, S.; Khan, A.; Mumtaz, S.U.; Chaudhry, M.M.; Bertuccioli, A.; Derosa, G.; Maffioli, P.; Togni, S.; Riva, A.; et al. Potential Clinical Benefits of Quercetin in the Early Stage of COVID-19: Results of a Second, Pilot, Randomized, Controlled and Open-Label Clinical Trial. *Int. J. Gen. Med.* **2021**, *14*, 2807–2816. [CrossRef]
50. Mangiavacchi, F.; Botwina, P.; Menichetti, E.; Bagnoli, L.; Rosati, O.; Marini, F.; Fonseca, S.; Abenante, L.; Alves, D.; Dabrowska, A.; et al. Seleno-Functionalization of Quercetin Improves the Non-Covalent Inhibition of Mpro and Its Antiviral Activity in Cells against SARS-CoV-2. *Int. J. Mol. Sci.* **2021**, *22*, 7048. [CrossRef]
51. Aucoin, M.; Cooley, K.; Saunders, P.R.; Cardozo, V.; Remy, D.; Cramer, H.; Abad, C.N.; Hannan, N. The effect of quercetin on the prevention or treatment of COVID-19 and other respiratory tract infections in humans: A rapid review. *Adv. Integr. Med.* **2020**, *7*, 247–251. [CrossRef]
52. Chaves, J.O.; de Souza, M.C.; Da Silva, L.C.; Perez, D.L.; Mayanga, P.C.T.; Machado, A.P.D.F.; Forster-Carneiro, T.; Vázquez-Espinosa, M.; González-De-Peredo, A.V.; Barbero, G.F.; et al. Extraction of Flavonoids From Natural Sources Using Modern Techniques. *Front. Chem.* **2020**, *8*, 507887. [CrossRef] [PubMed]
53. Nia, N.N.; Hadjmohammadi, M.R. The application of three-phase solvent bar microextraction based on a deep eutectic solvent coupled with high-performance liquid chromatography for the determination of flavonoids from vegetable and fruit juice samples. *Anal. Methods* **2019**, *11*, 5134–5141. [CrossRef]
54. Stefou, I.; Grigorakis, S.; Loupassaki, S.; Makris, D.P. Development of sodium propionate-based deep eutectic solvents for polyphenol extraction from onion solid wastes. *Clean Technol. Environ. Policy* **2019**, *21*, 1563–1574. [CrossRef]
55. Eugenio, M.H.A.; Pereira, R.G.F.A.; De Abreu, W.C.; Pereira, M.C.D.A. Phenolic compounds and antioxidant activity of tuberous root leaves. *Int. J. Food Prop.* **2017**, *20*, 2966–2973. [CrossRef]
56. Sharma, K.; Mahato, N.; Nile, S.H.; Lee, E.T.; Lee, Y.R. Economical and environmentally-friendly approaches for usage of onion (*Allium cepa* L.) waste. *Food Funct.* **2016**, *7*, 3354–3369. [CrossRef]
57. Puri, C.; Pucciarini, L.; Tiecco, M.; Brighenti, V.; Volpi, C.; Gargaro, M.; Germani, R.; Pellati, F.; Sardella, R.; Clementi, C. Use of a Zwitterionic Surfactant to Improve the Biofunctional Properties of Wool Dyed with an Onion (*Allium cepa* L.) Skin Extract. *Antioxidants* **2020**, *9*, 1055. [CrossRef] [PubMed]
58. Lee, S.Y.; Yim, D.G.; Hur, S.J. Changes in the Content and Bioavailability of Onion Quercetin and Grape Resveratrol During In Vitro Human Digestion. *Foods* **2020**, *9*, 694. [CrossRef]
59. Pucciarini, L.; Ianni, F.; Petesse, V.; Pellati, F.; Brighenti, V.; Volpi, C.; Gargaro, M.; Natalini, B.; Clementi, C.; Sardella, R. Onion (*Allium cepa* L.) Skin: A Rich Resource of Biomolecules for the Sustainable Production of Colored Biofunctional Textiles. *Molecules* **2019**, *24*, 634. [CrossRef]
60. Celano, R.; Docimo, T.; Piccinelli, A.; Gazzerro, P.; Tucci, M.; Di Sanzo, R.; Carabetta, S.; Campone, L.; Russo, M.; Rastrelli, L. Onion Peel: Turning a Food Waste into a Resource. *Antioxidants* **2021**, *10*, 304. [CrossRef]
61. Rashed, K.; Ciric, A.; Glamočlija, J.; Sokovic, M. Antibacterial and antifungal activities of methanol extract and phenolic compounds from *Diospyros virginiana* L. *Ind. Crop. Prod.* **2014**, *59*, 210–215. [CrossRef]
62. Hajlaoui, H.; Trabelsi, N.; Noumi, E.; Snoussi, M.; Fallah, H.; Ksouri, R.; Bakhrouf, A. Biological activities of the essential oils and methanol extract of tow cultivated mint species (Mentha longifolia and Mentha pulegium) used in the Tunisian folkloric medicine. *World J. Microbiol. Biotechnol.* **2009**, *25*, 2227–2238. [CrossRef]
63. Zang, Y.-Y.; Yang, X.; Chen, Z.-G.; Wu, T. One-pot preparation of quercetin using natural deep eutectic solvents. *Process Biochem.* **2020**, *89*, 193–198. [CrossRef]
64. ESmith, E.L.; Abbott, A.; Ryder, K. Deep Eutectic Solvents (DESs) and Their Applications. *Chem. Rev.* **2014**, *114*, 11060–11082. [CrossRef]
65. Shekaari, H.; Zafarani-Moattar, M.T.; Shayanfar, A.; Mokhtarpour, M. Effect of choline chloride/ethylene glycol or glycerol as deep eutectic solvents on the solubility and thermodynamic properties of acetaminophen. *J. Mol. Liq.* **2018**, *249*, 1222–1235. [CrossRef]

66. Cardellini, F.; Tiecco, M.; Germani, R.; Cardinali, G.; Corte, L.; Roscini, L.; Spreti, N. Novel zwitterionic deep eutectic solvents from trimethylglycine and carboxylic acids: Characterization of their properties and their toxicity. *RSC Adv.* **2014**, *4*, 55990–56002. [CrossRef]
67. Abraham, M.H.; Acree, W.E. On the solubility of quercetin. *J. Mol. Liq.* **2014**, *197*, 157–159. [CrossRef]
68. Singh, R.; Singh, S.; Kumar, S.; Arora, S. Studies on antioxidant potential of methanol extract/fractions of Acacia auriculiformis A. Cunn. *Food Chem.* **2007**, *103*, 505–511. [CrossRef]
69. Gutfinger, T. Polyphenols in olive oils. *J. Am. Oil Chem. Soc.* **1981**, *58*, 966–968. [CrossRef]
70. Ma, C.; Laaksonen, A.; Liu, C.; Lu, X.; Ji, X. The peculiar effect of water on ionic liquids and deep eutectic solvents. *Chem. Soc. Rev.* **2018**, *47*, 8685–8720. [CrossRef] [PubMed]
71. Hammond, O.S.; Bowron, D.T.; Edler, K.J. The Effect of Water upon Deep Eutectic Solvent Nanostructure: An Unusual Transition from Ionic Mixture to Aqueous Solution. *Angew. Chem. Int. Ed.* **2017**, *56*, 9782–9785. [CrossRef]
72. Kaur, S.; Malik, A.; Kashyap, H.K. Anatomy of Microscopic Structure of Ethaline Deep Eutectic Solvent Decoded through Molecular Dynamics Simulations. *J. Phys. Chem. B* **2019**, *123*, 8291–8299. [CrossRef] [PubMed]
73. Mocan, A.; Diuzheva, A.; Bădărău, S.; Moldovan, C.; Andruch, V.; Carradori, S.; Campestre, C.; Tartaglia, A.; De Simone, M.; Vodnar, D.; et al. Liquid Phase and Microwave-Assisted Extractions for Multicomponent Phenolic Pattern Determination of Five Romanian Galium Species Coupled with Bioassays. *Molecules* **2019**, *24*, 1226. [CrossRef]
74. Pérez-Gregorio, R.M.; García-Falcón, M.S.; Simal-Gándara, J.; Rodrigues, A.S.; Almeida, D.P. Identification and quantification of flavonoids in traditional cultivars of red and white onions at harvest. *J. Food Compos. Anal.* **2010**, *23*, 592–598. [CrossRef]
75. Ren, F.; Reilly, K.; Kerry, J.P.; Gaffney, M.; Hossain, M.; Rai, D.K. Higher Antioxidant Activity, Total Flavonols, and Specific Quercetin Glucosides in Two Different Onion (Allium cepa L.) Varieties Grown under Organic Production: Results from a 6-Year Field Study. *J. Agric. Food Chem.* **2017**, *65*, 5122–5132. [CrossRef]
76. Rhodes, M.J.; Price, K.R. Analytical problems in the study of flavonoid compounds in onions. *Food Chem.* **1996**, *57*, 113–117. [CrossRef]
77. Su, J.; Fu, P.; Shen, Y.; Zhang, C.; Liang, M.; Liu, R.; Li, H.; Zhang, W. Simultaneous analysis of flavonoids from Hypericum japonicum Thunb.ex Murray (Hypericaceae) by HPLC-DAD–ESI/MS. *J. Pharm. Biomed. Anal.* **2008**, *46*, 342–348. [CrossRef]
78. Wilm, M. Principles of Electrospray Ionization. *Mol. Cell. Proteom.* **2011**, *10*, M111.009407. [CrossRef]
79. Abdelkawy, K.S.; Balyshev, M.E.; Elbarbry, F. A new validated HPLC method for the determination of quercetin: Application to study pharmacokinetics in rats. *Biomed. Chromatogr.* **2017**, *31*, e3819. [CrossRef]
80. Pandey, J.; Bastola, T.; Tripathi, J.; Tripathi, M.; Rokaya, R.K.; Dhakal, B.; Rabin, D.C.; Bhandari, R.; Poudel, A. Estimation of Total Quercetin and Rutin Content in *Malus domestica* of Nepalese Origin by HPLC Method and Determination of Their Antioxidative Activity. *J. Food Qual.* **2020**, *2020*, 1–13. [CrossRef]
81. Savic, I.M.; Nikolic, V.D.; Nikolic, L.B.; Stankovic, M.Z. Development and validation of a new RP-HPLC method for determination of quercetin in green tea. *J. Anal. Chem.* **2013**, *68*, 906–911. [CrossRef]
82. Ruesgas-Ramón, M.; Figueroa-Espinoza, M.C.; Durand, E. Application of Deep Eutectic Solvents (DES) for Phenolic Compounds Extraction: Overview, Challenges, and Opportunities. *J. Agric. Food Chem.* **2017**, *65*, 3591–3601. [CrossRef]

Review

Promising Technological and Industrial Applications of Deep Eutectic Systems

Alberto Mannu *, Marco Blangetti, Salvatore Baldino and Cristina Prandi *

Department of Chemistry, University of Turin, Via Pietro Giuria 7, I-10125 Turin, Italy; marco.blangetti@unito.it (M.B.); salvatore.baldino@unito.it (S.B.)
* Correspondence: alberto.mannu@gmail.com (A.M.); cristina.prandi@unito.it (C.P.)

Abstract: Deep Eutectic Systems (DESs) are obtained by combining Hydrogen Bond Acceptors (HBAs) and Hydrogen Bond Donors (HBDs) in specific molar ratios. Since their first appearance in the literature in 2003, they have shown a wide range of applications, ranging from the selective extraction of biomass or metals to medicine, as well as from pollution control systems to catalytic active solvents and co-solvents. The very peculiar physical properties of DESs, such as the elevated density and viscosity, reduced conductivity, improved solvent ability and a peculiar optical behavior, can be exploited for engineering modular systems which cannot be obtained with other non-eutectic mixtures. In the present review, selected DESs research fields, as their use in materials synthesis, as solvents for volatile organic compounds, as ingredients in pharmaceutical formulations and as active solvents and cosolvents in organic synthesis, are reported and discussed in terms of application and future perspectives.

Keywords: deep eutectic solvents; materials; API formulation; gas sorbents; ionothermal synthesis; organic synthesis

Citation: Mannu, A.; Blangetti, M.; Baldino, S.; Prandi, C. Promising Technological and Industrial Applications of Deep Eutectic Systems. *Materials* **2021**, *14*, 2494. https://doi.org/10.3390/ma14102494

Academic Editor: Francisco Del Monte

Received: 29 March 2021
Accepted: 10 May 2021
Published: 12 May 2021

Publisher's Note: MDPI stays neutral with regard to jurisdictional claims in published maps and institutional affiliations.

Copyright: © 2021 by the authors. Licensee MDPI, Basel, Switzerland. This article is an open access article distributed under the terms and conditions of the Creative Commons Attribution (CC BY) license (https://creativecommons.org/licenses/by/4.0/).

1. Introduction

The so-called Deep Eutectic Solvents (or DESs) have been firstly described in the scientific literature in 2003 with the pioneering work of Abbot and co-workers on the eutectic behavior of urea in the presence of some quaternary ammonium salts [1].

Since this seminal report, many efforts have been devoted in the last decade to the development of such systems which revealed impressive performances when used as green solvents. In particular, their peculiar physical properties, such as low volatility, flammability and vapor pressure, increased chemical and thermal stability could be exploited for generating a superior class of improved solvents [2]. In addition, by designing DESs with specific components, it is possible to obtain eutectic mixtures with low toxicity and high biodegradability [3].

Generally speaking, a DES is obtained by combining in eutectic molar ratio specific Hydrogen bond Donors (HBDs) and Acceptors (HBAs). The most evident, as easily visible, parameter related with the formation of a DES is represented by its melting point, which results lower than the one of the formers HBDs and HBAs, typically with values around the room temperature. As an example, when the quaternary ammonium salt choline chloride is mixed with urea (both solid at room temperature) in a 1:2 molar ratio, a viscous liquid is obtained in a few minutes. When all the possible combinations between any HBA and HBD are considered, the definition of DES becomes tricky, especially in terms of distinction between Deep Eutectic (DES) and Eutectic Solvents (ES). This matter has been the subject of research and discussion for many years. Currently, according to Ryder [4] and Martins [5], the term *deep* should be associated with eutectic mixtures that show a decreasing of the melting point with respect to the ideal eutectic point and not with respect to the melting points of the former HBA and HBD. In the absence of such a condition, a simple eutectic mixture (ES) is thus obtained.

Regarding the possibility to classify a DES according to the nature of the formers HBA and HBD, Abbott and co-workers [6] defined Deep Eutectic Solvents using the general formula:

$$Cat^+X^- \cdot Y^- \quad (1)$$

where Cat^+ is an organic cation (typically an ammonium, but also phosphonium or sulphonium are included), X^- is the Lewis base counterion (generally a halide ion) and Y represents a Lewis or Brønsted acid which is involved in the formation of the anionic complex with X^-. Nowadays, the majority of DESs that have been prepared and studied could be classified into five different classes, as shown in Table 1.

Table 1. Classification of deep eutectic solvents.

DES Classification	General Formula	Terms
Type I	$Cat^+X^- \cdot MCl_x$	M = Zn, Sn, Fe, Al, Ga
Type II	$Cat^+X^- \cdot MCl_x \cdot yH_2O$	M = Cr, Co, Cu, Ni, Fe
Type III	$Cat^+X^- \cdot RZ$	Z = $CONH_2$, COOH, OH
Type IV	$MCl_x \cdot RZ$	M = Al, Zn; Z = $CONH_2$, COOH, OH
Type V	non ionic	

Type I and Type II DESs combine a quaternary ammonium salt and a metal chloride, the latter both in anhydrous (Type I) or in hydrated form (Type II). DESs can be formulated by the combination of a quaternary ammonium salt with a HBD, as in the above mentioned ChCl-based DES, where the HBD species is typically a small organic molecule (Type III DESs). Type IV, a combination between Type II and Type III DESs, includes the deep eutectic mixtures formed by a metal chloride hydrate and an organic HBD [6]. Recently, a fifth class of DES has been integrated in the general classification including all the deep eutectics composed of only non-ionic, molecular HBAs and HBDs (Type V DESs) [7].

Type III DESs are the most investigated class of DESs. The HBA is typically an ionic halide salt, such as an ammonium or phosphonium salt. Choline chloride (ChCl), a quaternary ammonium salt, is one of the most employed HBAs for the formation of DESs, since it fulfils several sustainability principles due to its reduced costs, high biodegradability, low toxicity and bioavailability. The other component of the DES is a generally safe and bioavailable small molecule as urea, organic carboxylic acids (e.g., mono- or bicarboxylic acids, citric acid or aminoacids) or polyols (e.g., glycerol, ethylene glycol or carbohydrates). Some selected examples of HBAs and HBDs are illustrated in Figure 1.

As a matter of fact, the almost unlimited possible combinations and the large pool of bio-derived and bio-inspired components available offer boundless possibilities to formulate new DESs with remarkable properties in terms of physico-chemical parameters. A relevant complementary class of DESs is represented by Natural Deep Eutectic Solvents (NADESs). The term NADES is generally intended to designate DESs composed only by naturally occurring compounds. In 2011 Verpoorte observed that a small number of primary metabolites as carboxylic acids, choline, sugars and aminoacids are present in high amounts in living organisms, much more abundant than expected on the basis of their metabolic roles [8]. It has also been proposed that NADESs may be involved in the resistance of some organisms to low temperatures and drought [9,10]. The hypothesis put forward to explain the ubiquitous presence of NADESs in living organisms is that the mixture of metabolites could form eutectic mixtures that would serve as reaction media for the biosynthesis of non-water-soluble molecules [8].

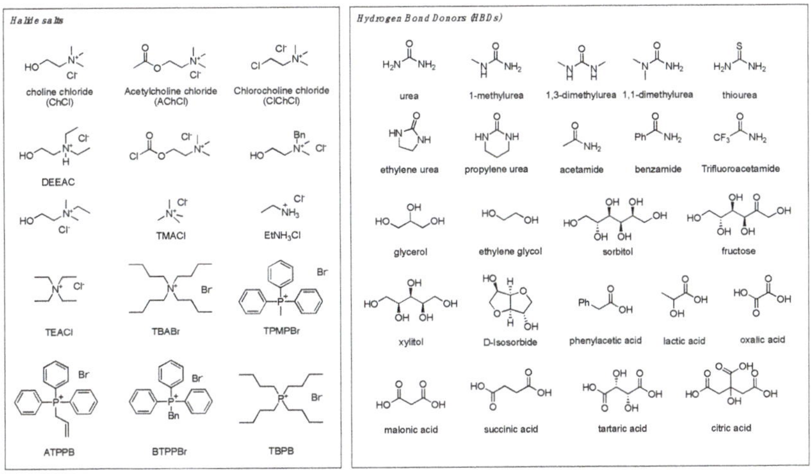

Figure 1. Selected examples of HBAs and HBDs for the formulation of DESs. Bn = benzyl, Ph = phenyl.

Taking inspiration from nature, several eutectic mixtures of bioderived compounds have been prepared and employed for various applications [11–13]. Among the wide variety in terms of components of NADESs, hydrophobic mixtures based on terpenes and fatty acids have been reported [14–17]. A growing number of publications report on very diverse use of NADESs. The special features of NADESs, such as biodegradability and biocompatibility [11], suggest that they are alternative candidates for concepts and applications and can be used in replacement to Volatile Organic Solvents (VOCs) in organic synthesis as well as in extraction processes [2,18,19]. Stemming from the idea that NADESs are the natural environment for metabolic processes, biocatalysis in NADESs has received much attention for those transformations in which substrates or products show low water solubility [20–22]. Most of the applications rely on the use of NADESs as extraction media for Natural Products (NPs). As NADES species exhibit a superior solubilizing ability for NPs, this provides a special advantage for NADESs as extraction media. Examples in the extraction capabilities of NADESs for a large number of natural products are covered in a pair of excellent recent reviews [23], and more recent references [24–28].

Further exploitation of NADES properties refers to their use as drug carriers. A recent study refers to the slightly reduced permeability of chloramphenicol through the pig skin after NADES application, which most likely was caused by hydrogen bonding between the NADES and proteins in the skin based on FT-IR-results [29]. Additional areas of possible further applications of NADES including their use as reaction, extraction and chromatographic media as well as their biomedical relevance are expected in the next future as far as the complexity of their supramolecular properties is elucidated.

The intriguing properties of DES make them exploitable not only as green solvents, but also in other fields of industrial interest. In this context, cutting edge applications of DESs are emerging, from the designing of functional materials [30], to the development of innovative medicinal formulations, to the possibility to act as active catalytic systems [31]. A relevant research line with important industrial applications is related with the employment of DESs as green solvents for biomass processing. This specific topic has been recently reviewed in a very exhaustive way by Jablonský and coworkers and thus will not be discussed herein [32].

In the present review, an overview about the application of DESs as possible organic liquid semiconductors, as shape directing agents, as solvents for VOCs and as additive or solvents in pharmaceutical formulations, is reported. In particular, a special focus on the molecular structure of these systems and the consequences on their performances will be given.

The present review has been divided into five parts.

In the first section, the relationship between the physical properties of DESs and their molecular structure will be discussed in general terms. Common preparation procedures will be also considered, as well as the low stability of some common DESs, recently highlighted by some Authors. In the second section, the application of eutectics in ionothermal synthesis and thus to the production of materials will be described. In the third part, the employment of DESs-based devices to the solubilization of Volatile Organic Compounds (VOCs) will be discussed. In the fourth section, promising application of some DESs in medicine (Therapeutic DESs, THEDESs), as in the formulation of improved drugs will be presented. In the last part, the possibility to employ DESs as non-innocent solvents in organic synthesis will be reported.

2. Molecular Structure of Deep Eutectic Systems and Properties

The so-called deep eutectic solvents (DESs) were initially considered as sort of room temperature ionic liquids [33]. Despite this first raw classification, DESs show many differences from the parent ionic liquids. DESs are not liquid salts, being the components very often neutral small molecules. They are in fact formed by eutectic combinations of HBAs and HBDs connected through an intense hydrogen bond network responsible for a very peculiar supramolecular architecture. In addition, the DESs' physical properties are uncommon as their enhanced solvent ability. Indeed, the possibility to produce sustainable DESs (as NADESs, by choosing appropriate HBAs and HBDs), make them potentially high performant green solvents. Nevertheless, during the last 15 years, an increasing number of alternative applications has been reported for many DESs, and some Authors started to use the term "DES" as short name for deep eutectic systems, in this way comprising any possible exploitation of such compounds. By the way, these eutectic systems have grown in importance as they show some peculiar physical properties directly related to their molecular structure. Several studies have been reported in order to characterize and classify many DESs considering some physical parameter, such as density, viscosity or conductivity. In addition, many experimental and theoretical attempts have been done in order to explain the DESs behavior at molecular level. A DES is usually characterized by a decreased viscosity, an increased density, a low conductivity and it usually shows liquid consistency at room temperature. In particular, when its melting point is plotted versus the molar fraction between the former constituents, a drop is experimentally observed in correspondence of the eutectic composition. When the drop is deeper than the expected theoretical melting point, the term DES results appropriately. Otherwise, the resulting mixture should be addressed as a simple ES (eutectic system) [4,34]. Many researchers studied the variation of physical properties as density and viscosity [35–38], refractive index [39] and speed of sound [40] of a DES and how these can be tailored by changing the nature of HBA, HBD or their composition. The possibility to engineer such systems makes of high priority the understanding of the supramolecular interactions which govern the formation of the DES.

In order to explain how the molecular structure and interactions are related to the experimentally observed melting point depression and to the specific physical characteristics commonly described, Abbot proposed the extension of the hole theory of Fürth to such systems [41]. In general, hole models are based on the statement that the vacancies present at molecular level are randomly distributed [42]. When an ionic mixture melts, the variation in temperature during the melting process produces fluctuations of the local density, which are responsible for the formation of empty spaces [43]. The consequent molecular framework is dynamic and allows, at some specific conditions, the constant movement of the ions with opportune size to fit in the holes, which move by this mechanism all-over the network. By exploiting the equations described by Abbot and co-workers, it is possible to determine the average size of the holes and to relate it with density and viscosity [9]. In particular, the relationship between the volume of the vacancies and the superficial tension, can be exploited to predict which HBAs and HBDs are more appropriated for customizing DESs with specific properties [44].

Recently, some attempts to compare the molecular structure of different DESs by meaning of their UV-VIS profile have been reported. Through the UV-VIS based Tauc plot method, the Band Gap (BG) energy of several hydrophilic choline-based [45] or hydrophobic phosphonium-based DESs [46] were determined and compared, revealing a relationship between the BG and the eutectic composition, which is analogue to the one observed when the melting point is expressed in terms of molar ratio between HBA and HBD (Table 2 and Figure 2).

Table 2. Band gap energies of some binary DESs.

HBA [1]	HBD [1]	Band Gap Energy (eV)	Reference
Choline chloride (1)	Glycolic acid (1)	4.67	[44]
Choline chloride (1)	Levulinic acid (1)	5.22	[44]
Choline chloride (1)	Ethylene glycol (2)	5.92	[44]
Choline acetate (1)	Glycolic acid (1)	4.73	[44]
Choline acetate (1)	Levulinic acid (1)	4.70	[44]
Choline acetate (1)	Ethylene glycol (2)	5.30	[44]
Choline chloride (1)	Zinc chloride (2)	5.78	[44]
Choline chloride (1)	Copper chloride (2)	5.20	[44]
Choline chloride (1)	Urea (2)	5.16	[44]
Choline chloride (1)	Nickel sulphate (1)	5.18	[45]
Choline chloride (3)	Imidazole (7)	4.74	[45]
Choline chloride (2)	D-(+)-Glucose (1)	5.85	[45]
Choline chloride (1)	Glycerol (5)	5.56	[45]
Triphenylmethylphosphonium bromide (1)	Ethylene glycol (5)	5.34	[45]
Triphenylmethylphosphonium bromide (1)	Glycerol (5)	5.23	[45]

[1] The molar equivalents of HBA and HBD are indicated in parenthesis.

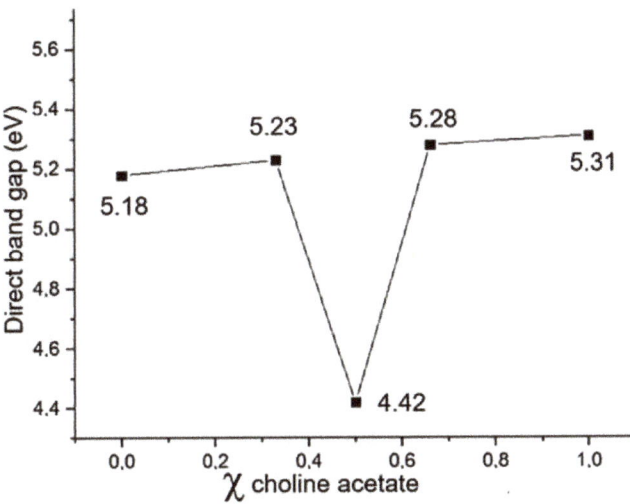

Figure 2. Variation of the band gap energy as function of the molar ratio between choline acetate and levulinic acid. Figure made from the data reported in reference [45].

The plot reported in Figure 2 represents the variation of the band gap energy as function of the molar fraction of choline acetate for the system choline acetate/levulinic acid. When the system contains equal moles of both the constituents, a drop of the bang gap energy is experimentally observed. It has been reported for several systems that the composition at which the band gap energy deeply decreases corresponds to the eutectic ratio between the formers HBAs and HBDs.

This UV-VIS-based procedure represents a fast methodology, alternative to the determination of the melting point, in order to find the correct eutectic composition of HBAs and HBDs mixtures. In addition, a relevant effect of the water on the BG energy was observed: the addition of 10% of water to choline based DESs resulted in a reduction of the BG energy. It is known that water addition on DESs reduces the lattice energy and this in part explains the trend observed. The possibility to tune some features of eutectic mixtures by adding specific amounts of water has been an object of intensive research activity, and many experimental and theoretical studies have been focused on the structural effect of water addition on eutectics [47]. It has been demonstrated that even low amounts of water affect sensibly the properties of a DES, but in order to produce a relevant structural variation, highly amount of water should be added. Hammond and co-workers [48] studied the water tolerance of the eutectic system composed by choline chloride and urea (1:2), revealing that only after a water addition superior to 42 wt% the system shows a transition from an ionic mixture to an aqueous solution, ceasing to exist in form of hydrated DES. At molecular level, it has been observed that the water tolerance is related to the DES nanostructure, which can host water molecules up to certain amounts without producing relevant structural changes [49].

Working below the limit of tolerance, it is possible to take advantage by the addition of water to a DES. One example of this can be represented by the choline chloride/urea (1:2) DES, which revealed to change its capacity to solubilize the CO_2 by adding even small amounts of water [50].

Even though the know-how about the specific effect of water content on the interaction between HBAs and HBDs is important for designing DESs with improved performances, the intrinsic hygroscopicity of eutectics is even more important. In fact, when DESs are used in industrial processes, the uptake of water from air cannot be avoided and, in this context, water is usually considered as an impurity [51]. Chen and co-workers determined the amount of water adsorbed (after 8 h) by a series of choline chloride based DESs containing as HBD glutaric acid, glycerol, ethylene glycol, xylitol, urea, glucose, methyl urea and oxalic acid. Most hygroscopic DESs (containing glycerol or ethylene glycol) showed not negligible adsorption values, in the range of 5–7 wt% [52]. As discussed before, these values are not high enough to disrupt the deep eutectic nature of the DES, nevertheless they must be taken into account in terms of influence on the DES physical properties. An exhaustive review paper concerning the effect of the water on specific DESs has been published in 2019 by El Achkar and co-workers [53].

As a matter of fact, the addition of water to a binary DES can be considered as the conversion of the original binary system in a new ternary one, where the water acts as a second HBD/HBA. This aspect is of general interest as the understanding of the combined effect of one HBAs and two HBDs on the properties of the resulting ionic mixture is pivotal for achieving a customization capacity. Recently, the effects of the preparation procedure, temperature and addition of a second HBD (water, methanol, 2-propanol, glycerol) were assessed for ternary mixtures containing choline chloride and ethylene glycol [54]. The Authors described an important effect of the time in the assessment of the DES structure after its preparation, indicating a fast ageing of these mixtures which should be considered for their application.

A further aspect of interest is related to the solvation of the anion in choline chloride based DESs. In fact, by ^{35}Cl NMR spectroscopy it is possible to describe the behavior of the anion in the presence of different amounts of water. Gabriele and coworkers correlated the ^{35}Cl linewidth with the increasing of weakness of DES-DES H-bond on samples of choline chloride/glycol (glycol = diethylene glycol, triethylene glycol, polyethylene glycol 200) upon addition of increasing amounts of water [55]. The same technology was employed by Di Pietro et al., who conducted ^{35}Cl NMR measurements on choline chloride based DESs (choline chloride/urea or glycolic acid) combined with theoretical studies and described the variation of the chloride solvation with the hydration of the system [56].

Each aspect related to the molecular structure of DESs and its consequences on their physical properties and thus on their possible applications should be contextualized in terms of operational window. In fact, DESs are not free from degradation, due to the simple fact that their components may decompose and/or react with each other. This possible issue, being reliable the purity of the single components, may basically occur in two moments: during and after the preparation of the eutectic mixture, depending, respectively, on the mixing and storage conditions employed. It has been reported that DESs usually display a thermal stability which is intermediate between their pure components [57–59], and HBDs are in general less thermally stable than the HBA. To the best of our knowledge, the possibility of a thermal degradation of DESs has been considered only for the systems ChCl/polyols and ChCl/carboxylic acids [60–62], whereas there are no reports on the thermal behavior of metal-based DESs.

The ChCl/carboxylic acid (e.g., oxalic, malic, malonic, levulinic acid) DESs appear to be the most sensitive systems compared with ChCl/polyols. As a matter of fact, all the evidence present in the literature has shown decomposition of the HBDs within the eutectic mixture even at relatively low temperatures (50–60 °C) [60,61]. The decomposition pathway observed consisted in the esterification between the HBD and ChCl, as inferred from ^{1}H NMR measurements, likely promoted by the acidic environment of the DES. Heating the mixtures at different temperatures (60–100 °C) for 2 h (Table 3) involves an increase of both the amount of ester product and water in the mixture (Scheme 1a), as measured by Karl-Fischer titration [61]. Apparently, the degradation process takes place regardless the preparation procedure employed, only at room temperature it occurs at much lower rate [60,61].

Table 3. Fraction of esterified ChCl in mol% after heating for 2 h at different temperatures estimated through ^{1}H NMR spectroscopy [61].

DES	ChCl Esterification (mol%)		
	60 °C	80 °C	100 °C
ChCl/lactic acid (1:2)	2	4	7
ChCl/levulinic acid (1:2)	2	10	17
ChCl/malic acid (1:2)	2	3	6
ChCl/oxalic acid (1:1)	0	6	17
ChCl/glutaric acid (1:1)	10	29	34
ChCl/malonic acid (1:1)	3	8	17

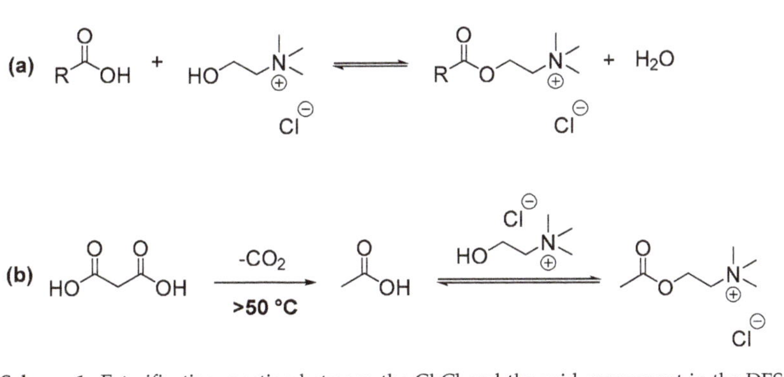

Scheme 1. Esterification reaction between the ChCl and the acid component in the DES (a) and parallel degradation pathway of MA in ChCl/MA (b).

In addition, DESs with a bicarboxylic acid component such as ChCl/malonic acid (MA) easily undergoes decarboxylation of MA to acetic acid, even occurring during its preparation if the operating temperature is above 50 °C, as demonstrated independently by Gontrani et al. [60] and Rodriguez Rodriguez et al. [61] This additional degradation pathway may increase the deterioration of the DES through a further esterification reaction between ChCl and acetic acid (Scheme 1b).

TGA studies demonstrated that degradation of ChCl/Gly (1:2) required elevated temperatures to occur [62]. FTIR data of the evolved gases suggest that over 200 °C CO_2, formaldehyde, acetaldehyde and water are formed as degradation products of Gly, possibly due to an intramolecular redox reaction under the harsh conditions (Scheme 2a). On the other hand, the ChCl constituent decomposed at higher temperatures (>300 °C), in line with the usual greater resistance of the HBA. ChCl likely degrades to chloromethane obtained by nucleophilic attack of the chloride anion on the methyl group of the ammonium moiety, liberating N,N-dimethylaminoethanol (Scheme 2b).

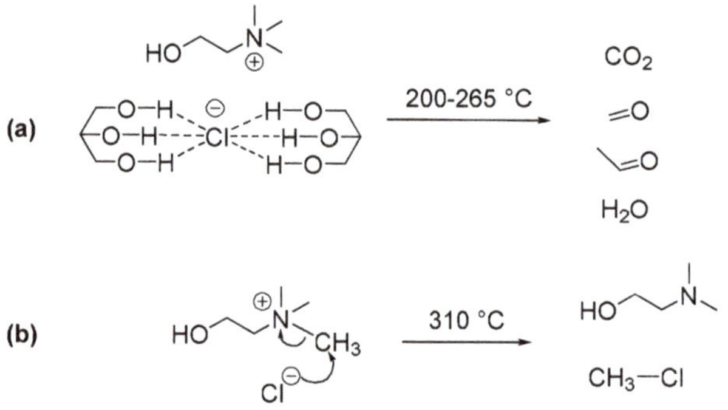

Scheme 2. Degradation of ChCl/Gly (1:2) under TGA conditions: decomposition products of Gly (**a**) and ChCl (**b**) (assessed by FTIR measurements).

3. DES for Ionothermal Synthesis

The know-how about the molecular structure of DESs can be exploited even for applications of industrial interest, as the synthesis of porous materials. Concerning this aspect, two subsequent key aspects should be considered: (I) the possibility to take advantage from the peculiar molecular structure of DES, (II) the chance to customize the DES structure by varying the nature of the HBA and HBD, and by modulating the preparation protocol.

In 2004 Cooper and co-workers reported an alternative protocol to the known hydrothermal synthesis for fabricating new zeotypes frameworks which was based on the unique properties on the DES formed by choline chloride and urea [63]. The Authors reported an innovative protocol for generate novel porous materials by exploiting the increased solvent ability of a DES combined with its well-established molecular structure, which resulted in being able to direct the synthesis toward specific spatial parameters. The possibility of using a DES not only as a solvent but also as Structure-Directing Agent (SDA) opened the way to several applied research lines. To date, many porous materials have been obtained in laboratory scale by exploiting specific combinations of HBA and HBD, and many review papers have well discussed the topic [64]. The process, named as ionothermal synthesis, can be performed with ionic liquids (ILs) or DESs and extended to the production of several porous materials (Table 4).

Table 4. Materials produced by ionothermal synthesis during the period 2016–2021.

Material	DES as Structure-Directing Agent
Imide-Linked Covalent Organic Frameworks	NaCl/KCl/ZnCl$_2$ [65]
Layered double hydroxides	Choline chloride/urea [66]
Fe-LEV aluminophosphate molecular sieves	Succinic acid/choline chloride/tetraethylammonium bromide [67]
Ti$_3$C$_2$ MXene [a]	Choline chloride and oxalic acid [68]
Triazine and heptazine polymeric carbon nitrides (PCNs)	NaCl/KCl [69]
MCM-41-supported metal catalysts	Choline chloride/glucose [70]
Fe$_3$O$_4$ magnetic nanoparticles	Choline chloride/urea [71]
Nanostructured ceria	Choline chloride/urea [72]
NEU20 [b]	Choline chloride/oxalic acid [73]
Cu-doped Fe$_3$O$_4$ nanoparticles	Choline chloride/urea [74]
gallium phosphate Ga$_3$(PO$_4$)$_4$(C$_2$N$_2$H$_8$)·(H$_2$C$_2$N$_2$H$_8$)$_2$·Cl	Choline chloride/imidazolinone [75]
NiCo$_2$O$_4$ Nanorods Decorated MoS$_2$ Nanosheets	Choline chloride/urea [76]
High-silica zeolites	Tetramethylammonium chloride/1,6-hexanediol Tetrapropylammonium bromide/pentaerythritol [77] Choline chloride/urea [78]

[a] Two-dimensional (2D) transition metal car- bides/nitrides from the 60+ group of MAX phases. [b] Photochromic inorganic–organic complex [C$_{10}$N$_2$H$_{10}$]$_2$[C$_{10}$N$_2$H$_8$][Ga$_2$(C$_2$O$_4$)$_5$].

4. Deep Eutectic Solvents for Gas Solubilization

The exclusive structure of DESs, which can be enriched with many functional groups by choosing opportune combinations of HBAs and HBDs, can be exploited to increase their gas sorption ability. This topic is of particular interest as Volatile Organic Compounds (VOCs) are common by-products in many industrial processes [79] and they are considered as hazard chemicals often associated with many diseases [80]. VOCs are produced in large amounts in the transport sector, and they are common ingredients of many cleaning products, representing one of the major sources of air pollutants [81,82]. Reduction of VOCs emission represents a priority, and it has been considered mandatory by several national and international normative [83].

From a technological point of view, trapping VOCs from gas streams with liquid sorbents represents a very efficient way for decreasing their presence in the environment.

The effectiveness of such an approach is related to two main aspects: (I) the availability of high performant sorbents, (II) the sustainability of the developed sorbents. Since a few years ago, the approaches employed to solve the above-mentioned problems have been revealed to be not efficient. The employment of organic solvents for VOCs removal cannot be pursued due to the toxicity and pollution associated with the sorbent devices [84,85]. On the other hand, the development of water and water-based sorbent devices would represent the best green alternative. Unfortunately, this possibility is precluded by the low solubility of the hydrophobic moieties contained in VOCs in the presence of water. More recently, the possibility to exploit ionic liquids for trapping VOCs has been explored with good results [86]. Nevertheless, high prices associated with the complex synthetic conditions of ionic liquids drastically reduce the overall sustainability of the process [87]. In this context, a possible advancement of the current available technology for VOCs treatment can be represented by the use of DESs. As discussed above, DESs are generally cheap, low pollutant, liquid at room temperature and they strongly interact with organic molecules. Theoretically, the design of highly efficient DESs- based VOCs sorbents constitutes the best technological solution to air pollution. Early studies about the employment of ammonium-based eutectics as solvent for the CO$_2$ were reported since 2015 (Li [88],

Mirza [89], Leron [90], Lu [91]) and the technology was extended to other VOCs two years later by Moura and co-workers [92].

In particular, Moura reported the possibility to use both hydrophilic choline chloride-based and hydrophobic tetrabutylphosphonium bromide-based DESs as adsorbents for toluene, acetaldehyde and dichloromethane (Table 5). Good results were reported, especially in the case of acetaldehyde which was adsorbed in percentages reaching the 99% with respect to the initial amount. Since these pioneering works, few extensions of the topic have been reported so far, especially in terms of VOC adsorbed.

Applications in SO_2 adsorption have been reported with quaternary ammonium salts-based DESs (HBD = glycerol [93], levulinic acid [94], guaiacol or cardanol [95]), with some thiocyanate-based DESs [96], and by employing eutectics formed by betaine or L-carnitine and ethylene glycol [97].

Regarding the application of DESs to the sorption of ammonia, a consistent family of DESs has been developed in the last years. Akhmetshina et al. reported a study on the gas sorbent properties of the DES 1-butyl-3-methyl imidazolium methanesulfonate/urea toward ammonia, hydrogen sulfide and carbon anhydride, with good results relative to the sorption capacity of the ammonia [98]. This study follows the results presented by Zhong [99,100] about phenol-based ternary DESs, by Yang using the hybrid DES choline chloride/resorcinol/glycerol (1:3:5) [101], Deng with protic NH_4SCN-based DESs [102] and Vorotyntsev who employed methanesulfonate-based DESs.

A different approach to the development of suitable DESs-based systems for trapping VOCs was reported by Di Pietro and co-workers [103]. The Authors demonstrated the possibility to change the ability of cyclodextrin to encapsulate toluene and aniline by performing the process in DES (choline chloride/urea 1:2). The reporting of this hybrid DES-cyclodextrin system opens to new possibilities of engineering by designing opportune eutectics and combining them with specific macromolecules.

Table 5 reports a resume of the main DESs employed as VOCs solvents and the corresponding adsorption capacity.

Table 5. Selected DESs employed as VOCs absorbents. The best adsorbing system in terms of temperature and pressure is herein reported.

VOC	DES [a]	Adsorption Capacity	Reference
toluene	ChCl/U (1:2)	2.8 wt% [b] (303 K)	[92]
	ChCl/EG (1:2)	2.5 wt% [b] (333 K)	[92]
	ChCl/GLY (1:2)	2.1 wt% [b] (303 K)	[92]
	ChCl/LA (1:2)	2.9 wt% [b] (303 K)	[92]
	TBPB/GLY (1:1)	2.9 wt% [b] (303 K)	[92]
	TBPB/LA (1:6)	3.0 wt% [b] (303 K)	[92]
	TBPB/DA (1:2)	3.0 wt% [b] (303 K)	[92]
acetaldehyde	ChCl/U (1:2)	2.9 wt% [b] (303 K)	[92]
	ChCl/EG (1:2)	2.9 wt% [b] (303 K)	[92]
	ChCl/GLY (1:2)	2.9 wt% [b] (303 K)	[92]
	ChCl/LA (1:2)	2.9 wt% [b] (303 K)	[92]
	TBPB/GLY (1:1)	2.9 wt% [b] (303 K)	[92]
	TBPB/LA (1:6)	3.0 wt% [b] (303 K)	[92]
	TBPB/DA (1:2)	3.0 wt% [b] (303 K)	[92]
CH_2Cl_2	ChCl /U (1:2)	2.0 wt% [b] (303 K)	[92]
	ChCl/EG (1:2)	2.6 wt% [b] (333 K)	[92]
	ChCl/GLY (1:2)	2.1 wt% [b] (303 K)	[92]
	ChCl /LA (1:2)	2.8 wt% [b] (303 K)	[92]
	TBPB/GLY (1:1)	2.8 wt% [b] (333 K)	[92]
	TBPB/LA (1:6)	2.8 wt% [b] (303 K)	[92]
	TBPB/DA (1:2)	3.0 wt% [b] (303 K)	[92]

Table 5. Cont.

VOC	DES [a]	Adsorption Capacity	Reference
CO$_2$	TBAB/TEA (1:5)	2.5 wt% [b] (303 K)	[88]
	TEAB/TEA (1:5)	2.9 wt% [b] (303 K)	[88]
	ChCl/MDEA (1:4)	3.0 wt% [b] (303 K)	[88]
	Ch/Cl/TEA (1:5)	3.0 wt% [b] (303 K)	[88]
	ChCl/MDEA (1:5)	4.0 wt% [b] (303 K)	[88]
	ChCl/TEA (1:4)	6.0 wt% [b] (303 K)	[88]
	ChCl/DEA (1:4)	15 wt% [b] (303 K)	[88]
	TBAB/MEA (1:5)	16 wt% [b] (303 K)	[88]
	TBAC/MEA (1:5)	17.5 wt% [b] (303 K)	[88]
	TEAB/MEA (1:5)	18 wt% [b] (303 K)	[88]
	TEAC/MEA (1:5)	22.5 wt% [b] (303 K)	[88]
	ChCl/MEA (1:4)	23 wt% [b] (303 K)	[88]
	TMAC/MEA (1:5)	23 wt% [b] (303 K)	[88]
	ChCl/MEA (1:5)	25 wt% [b] (303 K)	[88]
	TEAC/MEA/TEA	23 wt% [b] (303 K)	[88]
	ChCl/MEA/TEA	23 wt% [b] (303 K)	[88]
	TEAC/MEA/MDEA	23 wt% [b] (303 K)	[88]
	ChCl/MEA/MDEA	23 wt% [b] (303 K)	[88]
	TMAC/MEA/TEA	23 wt% [b] (303 K)	[88]
	TMAC/MEA/MDEA	30 wt% [b] (303 K)	[88]
	TMAC/MEA/FeCl$_3$ (1:5:0.1)	25 wt% [b] (303 K)	[88]
	TMAC/MEA/CuCl$_2$ (1:5:0.1)	26 wt% [b] (303 K)	[88]
	TMAC/MEA/NiCl$_2$ (1:5:0.1)	26 wt% [b] (303 K)	[88]
	TMAC/MEA/CoCl$_2$ (1:5:0.1)	26 wt% [b] (303 K)	[88]
	TMAC/MEA/NH$_4$Cl (1:5:0.1)	28 wt% [b] (303 K)	[88]
	TMAC/MEA/ZnCl$_2$ (1:5:0.1)	30 wt% [b] (303 K)	[88]
	TMAC/MEA/LiCl (1:5:0.1)	30 wt% [b] (303 K)	[88]
	ChCl/U (1:2)	3.559 mol$_{VOC}$/kg$_{DES}$ (303 K, 5.654 bar)	[90]
	1-butyl-3-methyl imidazolium methanesulfonate /U (1:1)	0.422 mol$_{VOC}$/kg$_{DES}$ (303 K, 6.984 bar)	[98]
NEt$_3$	ChCl/PhOH/EG (1:5:4)	9.619 mol$_{VOC}$/kg$_{DES}$ (298 K, 101 kPa)	[99]
	ChCl/PhOH/EG (1:7:4)	7.652 mol$_{VOC}$/kg$_{DES}$ (313 K, 101 kPa)	[99]
	ChCl/U (1:2)	2.213 mol$_{VOC}$/kg$_{DES}$ (298 K, 95 kPa)	[100]
	ChCl/resorcinol/GLY (1:3:5)	9.982 mol$_{VOC}$/kg$_{DES}$ (298 K, 101 kPa)	[101]
	ChCl/D-fructose/GLY (1:3:5)	6.471 mol$_{VOC}$/kg$_{DES}$ (313 K, 101 kPa)	[101]
	NH$_4$SCN/GLY (2:3)	10.353 mol$_{VOC}$/kg$_{DES}$ (298 K, 101 kPa)	[102]
SO$_2$	ChCl/guaiacol (1:3; 1:4; 1:5)	0.528; 0.501; 0.479 gVOC for gDES	[95]
	ChCl/cardanol (1:3; 1:4; 1:5)	0.196; 0.170; 0.149 gVOC for gDES	[95]
	ChCl/LA (1:3)	0.557 gVOC for gDES	[94]
	TBAC/LA (1:3)	0.622 gVOC for gDES	[94]
	ChCl/EG (1:2)	2.25 mol SO$_2$/mol DES	[104]
	ChCl/MA (1:1)	1.40 mol of SO$_2$ mol DES	[104]
	ChCl/U (1:2)	1.57 mol of SO$_2$ mol DES	[104]
	ChCl/ thiourea (1:1)	2.37 mol of SO$_2$ mol DES	[104]
NH$_3$	1-butyl-3-methyl imidazolium methanesulfonate /U (1:1)	4.150 mol$_{VOC}$/kg$_{DES}$ (313 K, 5.258 bar)	[98]

Table 5. Cont.

VOC	DES [a]	Adsorption Capacity	Reference
H_2S	1-butyl-3-methyl imidazolium methanesulfonate /U (1:1)	1.034 mol_{VOC}/kg_{DES} (303 K, 6.450 bar)	[98]

[a] ChCl: choline chloride, U: urea, GLY: glycerol, EG: ethylene glycol, LA: lactic acid, DA: decanoic acid, TBPB: tetrabutylphosphonium bromide, TBAB: tetrabutylammonium bromide, TEA: triethylamine, TEAB: tetraethylammonium bromide, MDEA: methyldiethanolamine, MEA: methylethanolamine, TEAC: tetraethylammonium chloride, TMAC: tetramethylammonium chloride. [b] Data extrapolated from the plots reported by the Authors. The numbers reported in the table are approximated.

5. Deep Eutectic Solvents in Medicine

One of the major challenges of the pharmaceutical industry is the improvement of existing active pharmaceutical ingredients (APIs) in terms of efficiency and pharmacological action, which are strongly correlated with a multitude of physico-chemical parameters such as solubility, permeation and bioavailability [105,106]. This approach is crucial in view of the development of new therapeutic agents, since it suppresses the clinical trial costs required in the drug development process.

In this context, the use of eutectic mixtures in the pharmaceutical field has a long-standing history and has found several applications both in drug delivery, whereas eutecticity improves drug solubility and permeability, and as reaction media in biocatalyzed reactions [107,108]. Recently, DESs and their derivatives have shown a great promising potential as drug delivery systems owing to their outstanding physico-chemical properties in terms of tunability, stability and low toxicology profiles [109,110]. In addition, some classes of DESs have been investigated as potential intrinsic therapeutic agents, showing a promising preliminary bioactivity *in vitro* against certain microorganisms and cancer cell lines. The applications of DESs as pharmaceutical tools have been a matter of a huge amount of research efforts over the last five years, and their improvement is currently under continuous development. The remarkable results achieved in this field have been recently extensively reviewed in the literature [111] and are summarized in Figure 3.

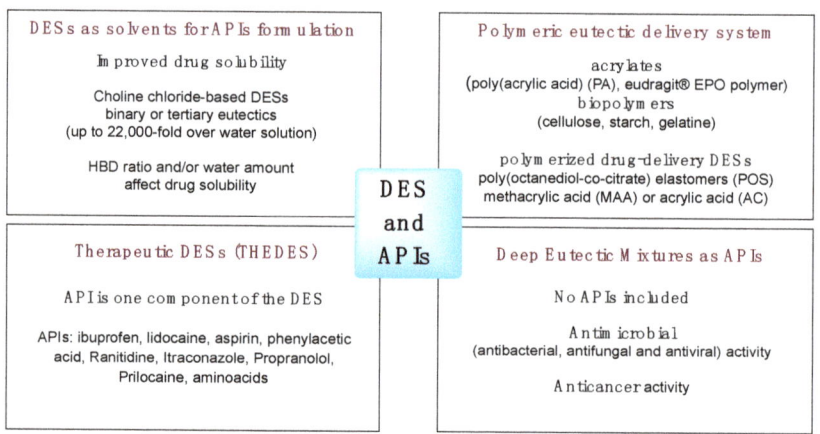

Figure 3. Pharmaceutical applications of Deep Eutectic Solvents: state-of-the-art.

DESs represent a safer and biocompatible alternative to organic solvents for the solubilization of poorly water-soluble APIs, in particular for topical formulations. Promising results on the solubilization of different classes of drug (nonsteroidal anti-inflammatory drugs, antifungal, anesthetics and analgesics) have been reported using mostly ChCl-based eutectic mixtures. While the HBA portion of these DESs (ChCl) is a safe and non-expensive compound which fulfils most of the sustainability principles, the choice of the HBDs has to be carefully addressed since many HBD components might possess a significant toxicity. Several drugs have shown a dramatically increase of their solubility in DESs compared to water, raised up to 5400-fold for ibuprofen [112], 6400-fold for posaconazole in binary eutectic systems [113] and up to 53,600-fold for the antifungal drug itraconazole in a ternary ChCl:Glycolic acid:oxalic acid (1:1.6:0.4) deep eutectic mixture [111]. Furthermore, DESs could improve the chemical stability of APIs, as observed for aspirin [110] and β-lactams antibiotics [114]. Several factors affect the solubility of APIs in DESs, among them the HBD ratios which can increase or decrease the solubility depending on the nature of both the drug and the eutectic, as a consequence of the different substrate-environment interactions [112]. Even the addition of an appropriate proportion of water to DESs or NADESs may significantly change their physicochemical properties and APIs solubility. As an example, the solubility of the benzylisoquinoline alkaloid berberine in a series of NADESs is enhanced up to 12-fold compared to water when switching from binary eutectics, where berberine solubility is lower than in water, to quaternary NADESs including water as a component [115].

Another powerful approach which has been extensively investigated to enhance APIs' solubility entails the incorporation of the active pharmaceutical species as a constituent of the deep eutectic mixture itself (API-DES or THEDES, Therapeutic Deep Eutectic Solvents) [116]. These eutectics can be designed using a variety of APIs (acting as HBAs or HBDs) and counterparts (e.g., metabolites) to fulfil specific therapeutic purposes [117]. API-DES formulations have shown remarkable results in the permeation enhancement of transdermal drug delivery systems. Several drugs such as ibuprofen [118], lidocaine [119] and itraconazole [120] incorporated with several permeation enhancers (terpenes or other drugs) in API-DES mixtures have shown both a remarkable solubility enhancement and an increased transdermal delivery in isotonic solution. API-DESs have been also exploited to improve some drug's oral bioavailability (e.g., CoQ10) [121], the intestinal absorption of daidzein [122] and the solubility and permeability of several drugs such as paeonol [123], ibuprofen and aspirin [112,124]. Moreover, the development of dual-drug eutectic systems incorporating two different APIs in the same eutectic formulation has opened the way to new fascinating strategies for synergic multimodal therapies using drugs with enhanced solubilization and permeation properties [125–127].

API-DES have been also exploited to control drug delivery as monomers for polymer production. These eutectic systems represent a significant advance in the development of controlled drug delivery systems, since they are able to (a) provide the API and (b) act both as monomer and reaction media for the polymerization reaction as a single formulation [128]. As an example, lidocaine has been incorporated in acrylic or methacrylic acid containing DESs which, after polymerization, allowed the controlled release of the anesthetic drug triggered by several parameters such as pH and ionic strength [129]. In addition, polymeric eutectic systems provide a simpler and greener alternative method for the incorporation of drugs and polymers. Several drug delivery systems have been investigated in (bio)polymer-based API-DES mixtures, among them anticancer (doxorubicin [130], Paclitaxel [131]), anti-inflammatory (ibuprofen [132], dexamethasone [133]) and anesthetic (prilocaine [134], lidocaine [135]) drugs, using poly(vinyl alcohol) (PV) and poly(acrylic acid) (PA) polymers, ammonium salts, SPCL (starch and poly-ε-caprolactone polymeric blend), cellulose [136], poly(octanediol-co-citrate) elastomers and gelatine [137] to tune the APIs release profiles.

In addition to drug delivery applications, DESs have recently shown promising pharmaceutical activities as antimicrobial (antiviral, antibacterial and antifungal) [138] and anticancer [139] agents. Preliminary studies in this field suggest that eutectic mixtures themselves have the potential to be further deeply investigated for the development of novel bio-inspired therapeutic agents.

6. Recent Advances in the Employment of DESs Both as Non-Innocent Solvents and as Active Co-Catalysts

The pursuit for the setup of sustainable processes in organic synthesis is a topic of paramount importance both from the academic and the industrial points of view. In this context, DESs have emerged as excellent solvents for environmentally benign reactions [4], especially in comparison with ILs, which have shown significant toxicity and extremely difficult preparation and purification procedures in several cases. Over the last fifteen years, DESs have extensively been used as reaction media for a large number of organic transformations, namely alkylation, condensation and multicomponent [18], organometallic reactions [140], together with sporadic bio- and transition metal-catalyzed processes [74]. DESs have also proven their feasibility as media in different types of processes, being employed for polymerization reactions [141], delignification of biomass feedstocks [142] and for the extraction [143] and purification of organic compounds from complex matrixes [144].

Among the impressive number of reports on the application of DESs in organic synthesis [18], those in which at least one component of the DES strongly participates to the transformation appear as the most appealing. The active role of the DES is achieved by reacting with other molecules present in the environment or by actively promoting the process, demonstrating a non-innocent effect of the eutectic mixture on the chemistry involved. After a careful examination of the literature, the examples that have shown a peculiar or a highly different impact of DESs with respect to VOCs on the same reaction can reasonably be grouped in three main classes: (i) polar organometallic chemistry; (ii) acid-mediated and (iii) transition-metal-catalyzed processes (Figure 4). The purpose of this section is to give an overview of the most interesting and recent advances in this scenario, which is continuously evolving, highlighting the employment and the great potential of these unconventional media as true protagonists and not only as mere spectators within the aforementioned areas.

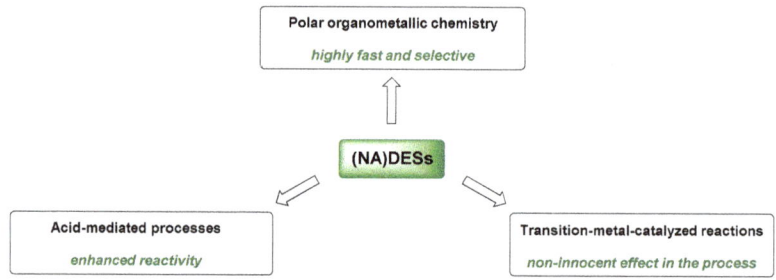

Figure 4. Fields of the most recent applications of DESs as solvents or co-catalysts in organic transformations.

6.1. Polar Organometallic Chemistry

The unravelling of the employment of highly reactive organometallic species, i.e., Grignard and organolithium (RLi) reagents in hydrophilic protic DESs is due to the seminal work of García-Álvarez's and Hevia's groups. They introduced the use of organomagnesium and RLi species in DESs for the reaction with ketones [145], non-activated imines [146] and for polymerization reaction to achieve synthetically relevant polyolefins [147], reaching high yields in several cases. All the reactions proceeded with high rate at 25–40 °C under air, giving in several cases improved yields and selectivity than standard protocols performed under inert atmosphere (Scheme 3).

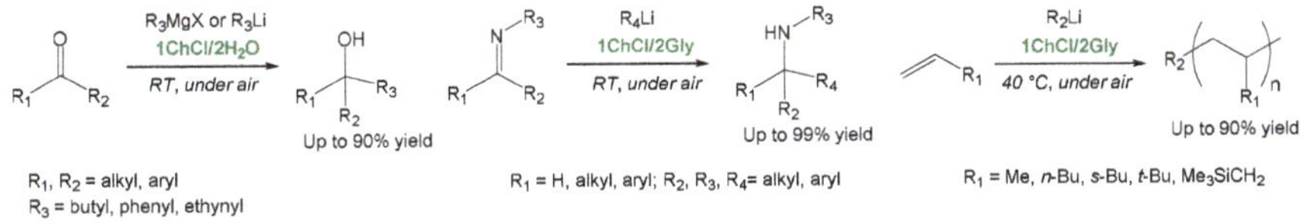

Scheme 3. First successful organic transformations featuring organometallic reagents in DESs under air.

Hence, it was demonstrated that the rate of addition of the organometallic species could successfully compete with their protonation by the DES medium, unlocking a new reactivity for the construction of several and diverse molecular frameworks.

Shortly afterwards, Mallardo et al. developed a methodology for the directed *ortho*-metalation (DoM) of 2,2-diphenyltetrahydrofuran (THF) using *t*-BuLi in the "greener" solvent CPME and sequential quenching with several electrophiles in ChCl/Gly (1:2) at 0 °C, gaining the *o*-substitution products chemoselectively at one phenyl ring with yields up to 90% within 10 min (Scheme 4) [148].

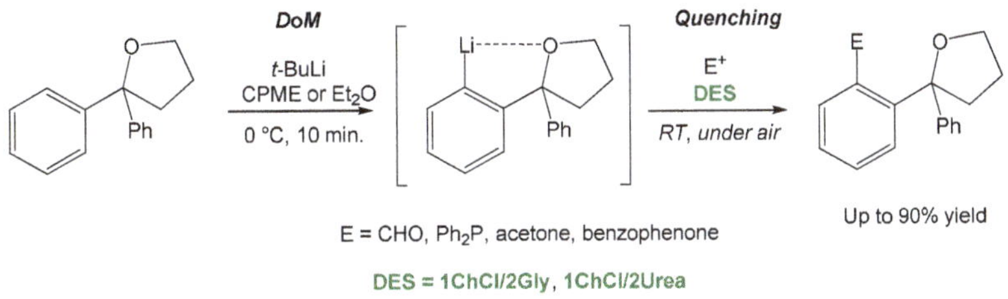

Scheme 4. Sequential DoM and quenching with electrophile for selective functionalization of 2,2-diphenylTHF.

Sassone et al. demonstrated that *o*-tolylTHF derivatives could undergo unprecedented alkylative ring-opening induced by directed lateral lithiation (DLL) in CPME/DES (ChCl/Gly (1:2)) mixture [149]. After quenching with several electrophiles, a functionalized primary alcohol is obtained (Scheme 5). It is worth pointing out that in the former paper the generation of the lithiated species was described in CPME, whereas in the latter the reactions occurred in a one-pot fashion. It appears that the addition of CPME to the DES mixture is required in order to stabilize the reactive RLi species over the competitive protonolysis process under these reaction conditions (Scheme 5).

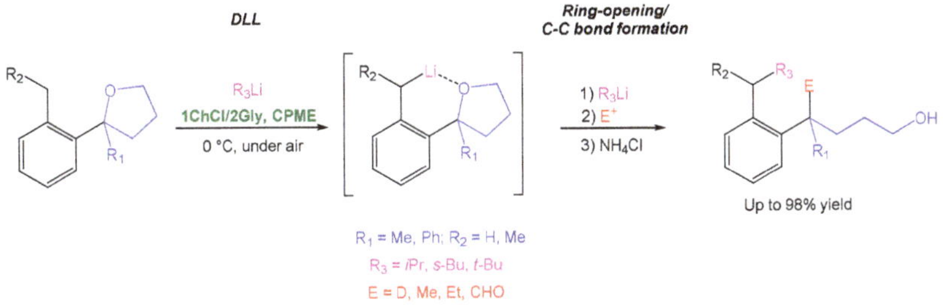

Scheme 5. One-pot DLL/ring-opening/C–C bond formation on *o*-tolylTHFs for the synthesis of functionalized primary alcohols.

The beneficial effect of CPME in this type of processes has also been observed by Ghinato et al. [150]. The authors described the reaction between the sterically hindered (hetero)arene *N,N*-diisopropylamides with RLi in CPME/DES (ChCl/Gly 1:2), easily modulating the chemoselectivity by changing the nature of the organolithium reagent. When R = *t*-Bu, ultrafast DoM process occurred and, after addition of the proper electrophile, functionalized amides were obtained (up to 95% yield), whereas using the less hindered MeLi, n-BuLi and n-HexLi, the nucleophilic character of the lithiated species prevailed, providing the corresponding ketones (up to 70% yield) via S_NAc (Scheme 6). Impressively, all the reactions occurred after 2 sec for the DoM process and within 1 min for the S_NAc reaction (0–25 °C). Interestingly, the use of pure VOCs or ChCl/Gly (1:2) led to a significant decrease in conversion and chemoselectivity, suggesting that the employment of a CPME/DES mixture is mandatory to achieve high yields of the desired products.

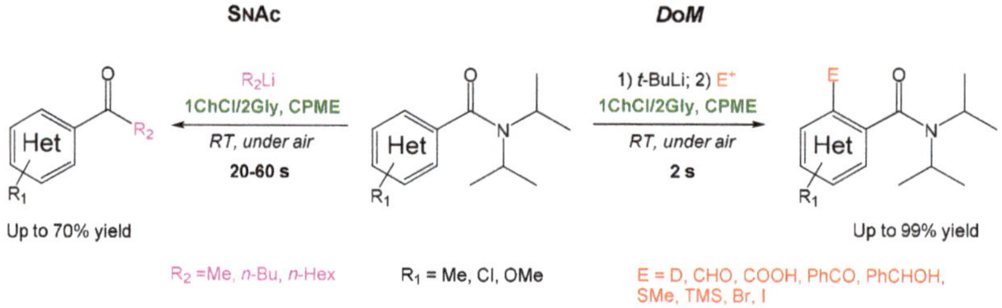

Scheme 6. Organolithium-directed chemoselectivity in ethereal ChCl/Gly (1:2): fast DoM and S_NAc on sterically hindered benzamides in CPME/DES.

The authors extended their methodology and investigated the lateral lithiation (LL) in DES on several o-tolyl-tertiary amides, sulfonamides and oxazolines, providing the DLL products functionalized with several electrophiles, over very short reaction times (Scheme 7) [151].

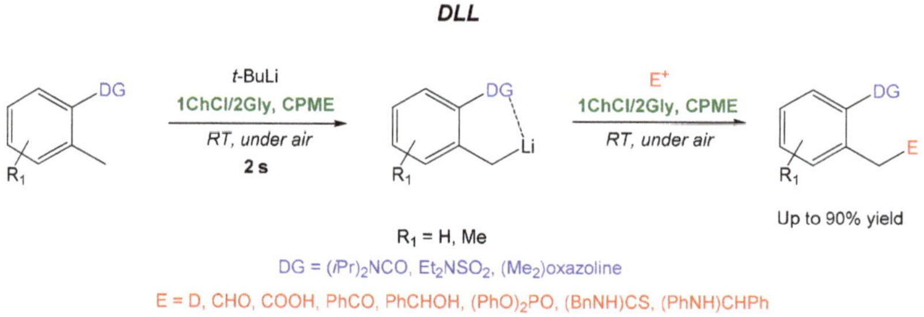

Scheme 7. Regioselective synthesis of toluene derivatives via ultrafast DLL in ethereal ChCl/Gly (1:2).

6.2. Acid-Mediated Reactions

Acid-based DESs have significantly been employed as media and reagents in different transformations, such as esterification [152], polymerization reactions [153] and for biomass valorization through the cleavage of chalcogenide moieties present in lignin feedstocks [154]. An interesting overview on the chemical and technological applications of Brønsted and Lewis acid based DESs, together with a discussion of their structural and acidity characteristics is offered by the recent survey of Qin et al. [155].

Among the acid-mediated processes, the Nazarov cyclization has arisen as a facile and highly atom economic route for the obtainment of cyclopentenone derivatives from divinyl ketone precursors. The process consists in a 4π conrotatory electrocyclization promoted by Brønsted or Lewis acids. Extensive studies employing (supported) organocatalysts and transition metal complexes have been performed [156], but only one example in DESs has been reported by Nejrotti et al. [157]. From their screening of several Brønsted-acid-based DESs, the authors found out ChCl/MA (malonic acid), ChCl/OA (oxalic acid) and ChCl/TsOH (p-toluenesulfonic acid) to be suitable system for promoting the reaction under mild conditions and in reasonable time (60 °C, 16 h). The scope was extended to complex molecular frameworks, accessing fused (hetero)polycyclic systems present in naturally occurring compounds containing the cyclopentenone motif (Scheme 8). Interestingly, the reported data indicated that carboxylic acids such as MA are not able to mediate the Nazarov cyclization in VOCs, hinting an increasing effect on the acidity of MA when it is a part of the DES mixture. This fact could be ascribed to its peculiar structure made of a thick network of hydrogen bonds created during the formation of the eutectic system.

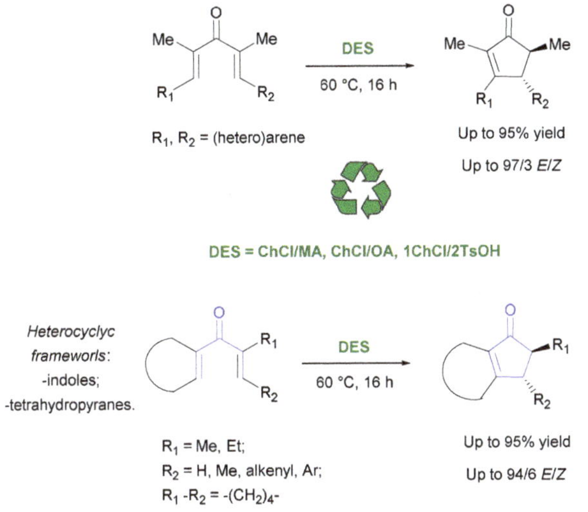

Scheme 8. Scope of the Nazarov cyclization in acidic DESs.

In addition, the reaction has proved to be performed on a gram-scale (E-factor = 19.5) and ChCl/MA has shown a good level of recyclability, mediating the process up to 4 runs with an overall acceptable yield (42%). The catalytic activity dramatically dropped at the 5th cycle, suggesting that the DES system could undergo decomposition processes throughout the reaction and the recycle conditions employed, likely decarboxylation of the acid component and trans-esterification between MA and the alcohol moiety of ChCl (*cf.* Section 1).

Very recently, the same authors reported the unprecedent Nazarov cyclization of a model divinyl ketone in two phosphonium-based DESs, (TPMPBr/EG (1:3) TPMPBr/AA (1:3)) using a two-level full factorial Design of Experiment approach for the optimization of the reaction conditions, in terms of reaction time, temperature and substrate concentration. In this case, the data indicate a strong cooperation between the two components of the DESs in promoting the Nazarov cyclization. Surface Responding Analysis (SRA) confirmed the synergic effect between the former components of the DES, which increases the expected performances of the system. Thus, conversions >90% with high chemoselectivity were reached under mild conditions (43 °C) [158].

6.3. Transition-Metal-Catalyzed Reactions

Lately, special attention has been devoted to the pursuit of sustainable catalytic reactions [159–162], and especially to transition-metal-mediated processes in DESs, with applications in Pd-catalyzed cross-coupling [163], Cu-catalyzed Ullmann type [164], Ru-catalyzed redox isomerization [165,166] and metathesis reactions [167]. Notwithstanding the increased sustainability in several cases, a distinct impact of the DES with respect to conventional systems (e.g., enhanced catalytic activity), is difficult to be addressed in the aforementioned examples. Particularly, two cases showing a strong participation of the DESs employed are worth mentioning and describing.

Cavallo et al. reported for the first time the Ru-catalyzed transfer hydrogenation of carbonyl compounds and imines using the DES both as reaction medium and H_2-source for the reduction process in the presence of TEA as the base under air [168]. After a thorough screening of different Ru-complexes and DESs, the most suitable system resulted TBABr/HCOOH, formic acid being the H_2-source in the presence of the diphosphane complex [RuCl$_2$(p-cymene)]$_2$-μ-dppf as the pre-catalyst (Scheme 9). Notwithstanding the highly reactive Ru-species involved in the transfer hydrogenation processes, no significant changes in the conversion were observed when the reaction was carried out under inert atmosphere. The recovery of the hydrogenated products was achieved after 4–16 h under mild conditions (40–60 °C) and the chemoselectivity generally ranged from moderate to excellent, showing clean crude mixtures in most cases.

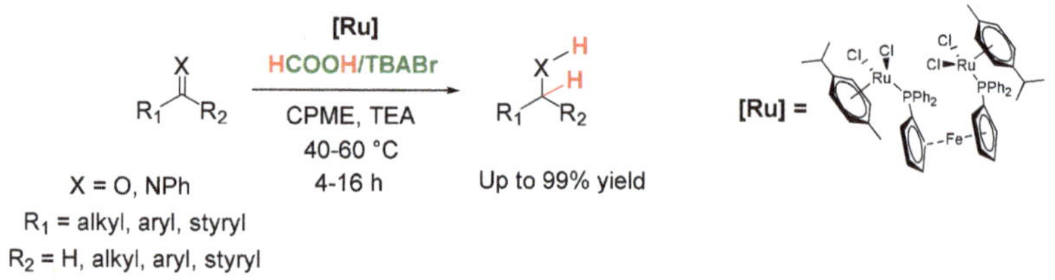

Scheme 9. Ru-catalyzed transfer hydrogenation of carbonyl and imine compounds in TBABr/HCOOH under mild conditions.

Interestingly, when the reduction of acetophenone was performed in VOCs (i.e., CPME, toluene, 2-MeTHF) with an equimolar mixture of HCOOH/NEt$_3$, the Ru-complex was not able to catalyze the process, as well as when a solution of TBABr and HCOOH was employed, even after several hours, indicating that the presence of the peculiar DES network was crucial for attaining the catalytic reduction (Scheme 9). However, the novel system does not appear to provide an easy recycling or regeneration procedure of the DES.

A bright example of a green transition-metal-catalyzed transformations has recently been reported by González-Sabín's and García-Álvarez's groups on the Meyer-Schuster rearrangement of propargylic alcohols [169]. In this work, the cheap eutectic mixture FeCl$_3$·6H$_2$O/Gly (3:1) successfully mediated the reaction of non-functionalized terminal and internal alkynols to their corresponding α,β-unsaturated carbonyl compounds with full conversions and yields up to 92%. The reactions proceeded very fast (5–30 min.) in the case of 1,1-diarylalkynols at room temperature, while the rearrangement required longer reaction times to occur (1–8 h) with 1,1-dialkylalkynols at a slightly higher temperature (40 °C). FeCl$_3$·6H$_2$O/Gly (3:1) could be easily recycled up to ten runs without appreciable loss of catalytic activity, showing a great stability of the system employed. It is noteworthy that the incorporation of FeCl$_3$·6H$_2$O within the reaction medium allows the recycling of both the catalyst and the solvent (Scheme 10).

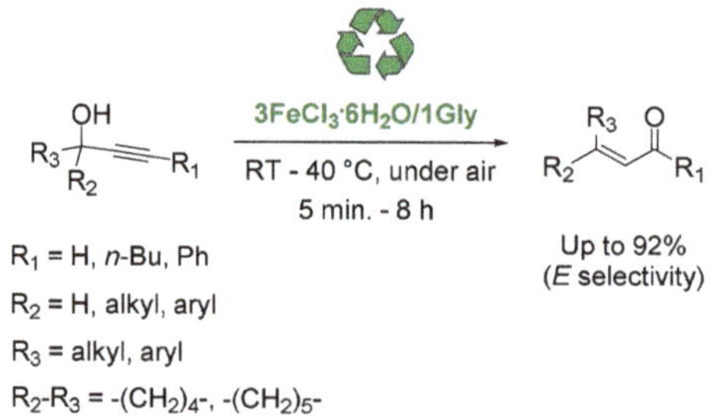

Scheme 10. Ligand-free iron salt incorporated within the DES: Meyer-Schuster rearrangement mediated by FeCl$_3$·6H$_2$O/Gly (3:1).

In addition, the successful although preliminary use of FeCl$_3$·6H$_2$O/Gly (3:1) in hydrolysis, cyclization and hydration reactions was also explored. It is worth pointing out that iron salts have scarcely been employed as catalysts for the Meyer-Schuster rearrangement in VOCs, but they required initiators or additives, higher temperatures (110–120 °C) and much longer reaction times (16 h). All the data demonstrate a reactivity enhancement of FeCl$_3$·6H$_2$O when part of the DES network with respect to its employment in conventional solvents. Notably, the mild reaction conditions, the facile recycle of the system and the use of a low-cost transition metal salt with no additional ligands in the place of expensive complexes confer to the process green credentials for sustainability.

7. Conclusions

During the last 20 years DESs have characterized a huge part of the overall research activity. The possibilities offered by the almost infinite number of eutectic combinations between HBAs and HBDs opened to the development of engineerable and modular systems. This fact, combined with the superior physical properties with respect to ILs, produced an intensive exploration of such systems in many research and applied fields, including many industrial applications. Some strategic fields of industrial interest have been herein discussed. The high levels of performance reached in ionothermal synthesis by using DESs both as solvents and as shape-directing agents are remarkable and open to the synthesis of new materials. In addition, the environmental impact of such systems when employed in VOCs treating devices is outstanding. In addition, the wide utilization of DESs in the pharmaceutical industry revealed their great promising potential as drug delivery systems to improve the pharmacokinetic properties of APIs and, in some cases, to act as APIs themselves. As a matter of fact, the exploitation of DESs is just in its early stages. In fact, as the understanding of their behavior at molecular level increases, new possibilities arise. The dual role of eutectic mixtures, as solvent and co-catalyst, together with their easy recyclability, observed in transition metal- and acid-mediated transformations could allow to upgrade many industrial processes in terms of efficiency and sustainability. Finally, the possibility to tune the structural disorder of mixtures of hydrogen bond acceptors and donors could give a relevant contribution to the development of new liquid organic semiconductors, with a cascade of industrial application currently incalculable. Regarding future outlooks, we can expect that DESs exploitation will take two main roads in the next years. From one side the relevant know-how available will be exploited for developing high effective systems in the fields of biomass conversion and in medicinal chemistry. From the other side, the extending of the eutectic concept not only to melting point but also to other physical features (as the band gap or Urbach energies) will open new application in the fields of optoelectronic, for the developing of organic liquid semiconductors of for the doping of existing systems. In addition, additional structure-activity studies are needed in order to understand how tailoring a specific hydrogen-bond network with the aim to direct the behavior of a chemical system, as an organic transformation or a catalytic process.

Author Contributions: Conceptualization, A.M. and C.P.; writing—original draft preparation, A.M.; M.B.; S.B. and C.P.; writing—review and editing, A.M.; M.B.; S.B. and C.P.; funding acquisition, C.P. All authors have read and agreed to the published version of the manuscript.

Funding: This research was funded by Italian Ministry of Research, Huvepharma Italia srl.

Institutional Review Board Statement: Not applicable.

Informed Consent Statement: Not applicable.

Data Availability Statement: No additional data available.

Acknowledgments: The authors would like to acknowledge Regione Piemonte for support.

Conflicts of Interest: The authors declare no conflict of interest. The funders had no role in the design of the study; in the collection, analyses or interpretation of data; in the writing of the manuscript or in the decision to publish the results.

References

1. Abbott, A.P.; Capper, G.; Davies, D.L.; Rasheed, R.K.; Tambyrajah, V. Novel solvent properties of choline chloride/urea mixtures. *Chem. Commun.* **2003**, 70–71. [CrossRef]
2. Perna, F.M.; Vitale, P.; Capriati, V. Deep eutectic solvents and their applications as green solvents. *Curr. Opin. Green Sustain. Chem.* **2020**, *21*, 27–33. [CrossRef]
3. Zhang, Q.; De Oliveira Vigier, K.; Royer, S.; Jérôme, F. Deep eutectic solvents: Syntheses, properties and applications. *Chem. Soc. Rev.* **2012**, *41*, 7108–7146. [CrossRef] [PubMed]
4. Smith, E.L.; Abbott, A.P.; Ryder, K.S. Deep Eutectic Solvents (DESs) and Their Applications. *Chem. Rev.* **2014**, *114*, 11060–11082. [CrossRef] [PubMed]
5. Martins, M.A.R.; Pinho, S.P.; Coutinho, J.A.P. Insights into the Nature of Eutectic and Deep Eutectic Mixtures. *J. Solut. Chem.* **2019**, *48*, 962–982. [CrossRef]
6. Abbott, A.P.; Barron, J.C.; Ryder, K.S.; Wilson, D. Eutectic-Based Ionic Liquids with Metal-Containing Anions and Cations. *Chem. Eur. J.* **2007**, *13*, 6495–6501. [CrossRef] [PubMed]
7. Abranches, D.O.; Martins, M.A.R.; Silva, L.P.; Schaeffer, N.; Pinho, S.P.; Coutinho, J.A.P. Phenolic hydrogen bond donors in the formation of non-ionic deep eutectic solvents: The quest for type V DES. *Chem. Commun.* **2019**, *55*, 10253–10256. [CrossRef] [PubMed]
8. Choi, Y.H.; van Spronsen, J.; Dai, Y.; Verberne, M.; Hollmann, F.; Arends, I.W.C.E.; Witkamp, G.-J.; Verpoorte, R. Are Natural Deep Eutectic Solvents the Missing Link in Understanding Cellular Metabolism and Physiology? *Plant Physiol.* **2011**, *156*, 1701–1705. [CrossRef]
9. Gertrudes, A.; Craveiro, R.; Eltayari, Z.; Reis, R.L.; Paiva, A.; Duarte, A.R.C. How Do Animals Survive Extreme Temperature Amplitudes? The Role of Natural Deep Eutectic Solvents. *ACS Sustain. Chem. Eng.* **2017**, *5*, 9542–9553. [CrossRef]
10. Craveiro, R.; Castro, V.I.B.; Viciosa, M.T.; Dionísio, M.; Reis, R.L.; Duarte, A.R.C.; Paiva, A. Influence of natural deep eutectic systems in water thermal behavior and their applications in cryopreservation. *J. Mol. Liq.* **2021**, *329*, 115533. [CrossRef]
11. Paiva, A.; Craveiro, R.; Aroso, I.; Martins, M.; Reis, R.L.; Duarte, A.R.C. Natural Deep Eutectic Solvents–Solvents for the 21st Century. *ACS Sustain. Chem. Eng.* **2014**, *2*, 1063–1071. [CrossRef]
12. Vanda, H.; Dai, Y.; Wilson, E.G.; Verpoorte, R.; Choi, Y.H. Green solvents from ionic liquids and deep eutectic solvents to natural deep eutectic solvents. *C. R. Chim.* **2018**, *21*, 628–638. [CrossRef]
13. Liu, Y.; Friesen, J.B.; McAlpine, J.B.; Lankin, D.C.; Chen, S.-N.; Pauli, G.F. Natural Deep Eutectic Solvents: Properties, Applications, and Perspectives. *J. Nat. Prod.* **2018**, *81*, 679–690. [CrossRef] [PubMed]
14. van Osch, D.J.G.P.; Dietz, C.H.J.T.; van Spronsen, J.; Kroon, M.C.; Gallucci, F.; van Sint Annaland, M.; Tuinier, R. A Search for Natural Hydrophobic Deep Eutectic Solvents Based on Natural Components. *ACS Sustain. Chem. Eng.* **2019**, *7*, 2933–2942. [CrossRef]
15. Florindo, C.; Branco, L.C.; Marrucho, I.M. Quest for Green-Solvent Design: From Hydrophilic to Hydrophobic (Deep) Eutectic Solvents. *ChemSusChem* **2019**, *12*, 1549–1559. [CrossRef] [PubMed]
16. Ribeiro, B.D.; Florindo, C.; Iff, L.C.; Coelho, M.A.Z.; Marrucho, I.M. Menthol-based Eutectic Mixtures: Hydrophobic Low Viscosity Solvents. *ACS Sustain. Chem. Eng.* **2015**, *3*, 2469–2477. [CrossRef]
17. Schaeffer, N.; Martins, M.A.R.; Neves, C.M.S.S.; Pinho, S.P.; Coutinho, J.A.P. Sustainable hydrophobic terpene-based eutectic solvents for the extraction and separation of metals. *Chem. Commun.* **2018**, *54*, 8104–8107. [CrossRef]
18. Alonso, D.A.; Baeza, A.; Chinchilla, R.; Guillena, G.; Pastor, I.M.; Ramón, D.J. Deep Eutectic Solvents: The Organic Reaction Medium of the Century. *Eur. J. Org. Chem.* **2016**, *2016*, 612–632. [CrossRef]
19. Cunha, S.C.; Fernandes, J.O. Extraction techniques with deep eutectic solvents. *TrAC Trends Anal. Chem.* **2018**, *105*, 225–239. [CrossRef]
20. Aranda, C.; de Gonzalo, G. Biocatalyzed Redox Processes Employing Green Reaction Media. *Molecules* **2020**, *25*, 3016. [CrossRef]
21. Guajardo, N.; Domínguez de María, P. Continuous Biocatalysis in Environmentally-Friendly Media: A Triple Synergy for Future Sustainable Processes. *ChemCatChem* **2019**, *11*, 3128–3137. [CrossRef]
22. Panić, M.; Radović, M.; Maros, I.; Jurinjak Tušek, A.; Cvjetko Bubalo, M.; Radojčić Redovniković, I. Development of environmentally friendly lipase-catalysed kinetic resolution of (R,S)-1-phenylethyl acetate using aqueous natural deep eutectic solvents. *Process Biochem.* **2021**, *102*, 1–9. [CrossRef]
23. Ruesgas-Ramón, M.; Figueroa-Espinoza, M.C.; Durand, E. Application of Deep Eutectic Solvents (DES) for Phenolic Compounds Extraction: Overview, Challenges, and Opportunities. *J. Agric. Food Chem.* **2017**, *65*, 3591–3601. [CrossRef]
24. Tong, X.; Yang, J.; Zhao, Y.; Wan, H.; He, Y.; Zhang, L.; Wan, H.; Li, C. Greener extraction process and enhanced in vivo bioavailability of bioactive components from Carthamus tinctorius L. by natural deep eutectic solvents. *Food Chem.* **2021**, *348*, 129090. [CrossRef]
25. Chandra Roy, V.; Ho, T.C.; Lee, H.-J.; Park, J.-S.; Nam, S.Y.; Lee, H.; Getachew, A.T.; Chun, B.-S. Extraction of astaxanthin using ultrasound-assisted natural deep eutectic solvents from shrimp wastes and its application in bioactive films. *J. Clean. Prod.* **2021**, *284*, 125417. [CrossRef]
26. Chen, M.; Lahaye, M. Natural deep eutectic solvents pretreatment as an aid for pectin extraction from apple pomace. *Food Hydrocoll.* **2021**, *115*, 106601. [CrossRef]

27. Cui, Z.; Enjome Djocki, A.V.; Yao, J.; Wu, Q.; Zhang, D.; Nan, S.; Gao, J.; Li, C. COSMO-SAC-supported evaluation of natural deep eutectic solvents for the extraction of tea polyphenols and process optimization. *J. Mol. Liq.* **2021**, *328*, 115406. [CrossRef]
28. Alrugaibah, M.; Yagiz, Y.; Gu, L. Use natural deep eutectic solvents as efficient green reagents to extract procyanidins and anthocyanins from cranberry pomace and predictive modeling by RSM and artificial neural networking. *Sep. Purif. Technol.* **2021**, *255*, 117720. [CrossRef]
29. Nystedt, H.L.; Grønlien, K.G.; Tønnesen, H.H. Interactions of natural deep eutectic solvents (NADES) with artificial and natural membranes. *J. Mol. Liq.* **2021**, *328*, 115452. [CrossRef]
30. Tomé, L.I.N.; Baião, V.; da Silva, W.; Brett, C.M.A. Deep eutectic solvents for the production and application of new materials. *Appl. Mater. Today* **2018**, *10*, 30–50. [CrossRef]
31. Ünlü, A.E.; Arıkaya, A.; Takaç, S. Use of deep eutectic solvents as catalyst: A mini-review. *Green Process. Synth.* **2019**, *8*, 355–372. [CrossRef]
32. Jablonský, M.; Škulcová, A.; Malvis, A.; Šima, J. Extraction of value-added components from food industry based and agro-forest biowastes by deep eutectic solvents. *J. Biotechnol.* **2018**, *282*, 46–66. [CrossRef] [PubMed]
33. Marsh, K.N.; Boxall, J.A.; Lichtenthaler, R. Room temperature ionic liquids and their mixtures—A review. *Fluid Phase Equilib.* **2004**, *219*, 93–98. [CrossRef]
34. Hammond, O.S.; Bowron, D.T.; Edler, K.J. Liquid structure of the choline chloride-urea deep eutectic solvent (reline) from neutron diffraction and atomistic modelling. *Green Chem.* **2016**, *18*, 2736–2744. [CrossRef]
35. Harifi-Mood, A.R.; Buchner, R. Density, viscosity, and conductivity of choline chloride+ethylene glycol as a deep eutectic solvent and its binary mixtures with dimethyl sulfoxide. *J. Mol. Liq.* **2017**, *225*, 689–695. [CrossRef]
36. Shekaari, H.; Zafarani-Moattar, M.T.; Mohammadi, B. Thermophysical characterization of aqueous deep eutectic solvent (choline chloride/urea) solutions in full ranges of concentration at T=(293.15–323.15)K. *J. Mol. Liq.* **2017**, *243*, 451–461. [CrossRef]
37. Ghaedi, H.; Ayoub, M.; Sufian, S.; Shariff, A.M.; Murshid, G.; Hailegiorgis, S.M.; Khan, S.N. Density, excess and limiting properties of (water and deep eutectic solvent) systems at temperatures from 293.15K to 343.15K. *J. Mol. Liq.* **2017**, *248*, 378–390. [CrossRef]
38. Ghaedi, H.; Ayoub, M.; Sufian, S.; Hailegiorgis, S.M.; Murshid, G.; Farrukh, S.; Khan, S.N. Experimental and prediction of volumetric properties of aqueous solution of (allyltriphenylPhosphonium bromide—Triethylene glycol) deep eutectic solvents. *Thermochim. Acta* **2017**, *657*, 123–133. [CrossRef]
39. Sánchez, P.B.; González, B.; Salgado, J.; José Parajó, J.; Domínguez, Á. Physical properties of seven deep eutectic solvents based on l-proline or betaine. *J. Chem. Thermodyn.* **2019**, *131*, 517–523. [CrossRef]
40. Leron, R.B.; Wong, D.S.H.; Li, M.-H. Densities of a deep eutectic solvent based on choline chloride and glycerol and its aqueous mixtures at elevated pressures. *Fluid Phase Equilib.* **2012**, *335*, 32–38. [CrossRef]
41. Abbott, A.P. Application of Hole Theory to the Viscosity of Ionic and Molecular Liquids. *ChemPhysChem* **2004**, *5*, 1242–1246. [CrossRef]
42. Stillinger, F.H., Jr.; Blander, M. (Eds.) *Molten Salt Chemistry*; Wiley Interscience: New York, NY, USA, 1964.
43. Fürth, R. On the theory of the liquid state: III. The hole theory of the viscous flow of liquids. *Proc. Camb. Phil. Soc.* **1941**, *37*, 281–290. [CrossRef]
44. Abbott, A.P.; Capper, G.; Gray, S. Design of Improved Deep Eutectic Solvents Using Hole Theory. *ChemPhysChem* **2006**, *7*, 803–806. [CrossRef] [PubMed]
45. Mannu, A.; Ferro, M.; Colombo Dugoni, G.; Di Pietro, M.E.; Garroni, S.; Mele, A. From deep eutectic solvents to deep band gap systems. *J. Mol. Liq.* **2020**, *301*, 112441. [CrossRef]
46. Mannu, A.; Di Pietro, M.E.; Mele, A. Band-Gap Energies of Choline Chloride and Triphenylmethylphosphoniumbromide-Based Systems. *Molecules* **2020**, *25*, 1495. [CrossRef] [PubMed]
47. Kaur, S.; Gupta, A.; Kashyap, H.K. Nanoscale Spatial Heterogeneity in Deep Eutectic Solvents. *J. Phys. Chem. B* **2016**, *120*, 6712–6720. [CrossRef] [PubMed]
48. Hammond, O.S.; Bowron, D.T.; Edler, K.J. The Effect of Water upon Deep Eutectic Solvent Nanostructure: An Unusual Transition from Ionic Mixture to Aqueous Solution. *Angew. Chem. Int. Ed.* **2017**, *56*, 9782–9785. [CrossRef]
49. Hammond, O.S.; Bowron, D.T.; Jackson, A.J.; Arnold, T.; Sanchez-Fernandez, A.; Tsapatsaris, N.; Garcia Sakai, V.; Edler, K.J. Resilience of Malic Acid Natural Deep Eutectic Solvent Nanostructure to Solidification and Hydration. *J. Phys. Chem. B* **2017**, *121*, 7473–7483. [CrossRef] [PubMed]
50. Su, W.C.; Wong, D.S.H.; Li, M.H. Effect of Water on Solubility of Carbon Dioxide in (Aminomethanamide + 2-Hydroxy-N,N,N-trimethylethanaminium Chloride). *J. Chem. Eng. Data* **2009**, *54*, 1951–1955. [CrossRef]
51. Ma, C.; Laaksonen, A.; Liu, C.; Lu, X.; Ji, X. The peculiar effect of water on ionic liquids and deep eutectic solvents. *Chem. Soc. Rev.* **2018**, *47*, 8685–8720. [CrossRef]
52. Chen, Y.; Yu, D.; Chen, W.; Fu, L.; Mu, T. Water absorption by deep eutectic solvents. *Phys. Chem. Chem. Phys.* **2019**, *21*, 2601–2610. [CrossRef]
53. El Achkar, T.; Fourmentin, S.; Greige-Gerges, H. Deep eutectic solvents: An overview on their interactions with water and biochemical compounds. *J. Mol. Liq.* **2019**, *288*, 111028. [CrossRef]
54. Mannu, A.; Cardano, F.; Fin, A.; Baldino, S.; Prandi, C. Choline chloride-based ternary deep band gap systems. *J. Mol. Liq.* **2021**, *330*, 115717. [CrossRef]

55. Gabriele, F.; Chiarini, M.; Germani, R.; Tiecco, M.; Spreti, N. Effect of water addition on choline chloride/glycol deep eutectic solvents: Characterization of their structural and physicochemical properties. *J. Mol. Liq.* **2019**, *291*, 111301. [CrossRef]
56. Di Pietro, M.E.; Hammond, O.; van den Bruinhorst, A.; Mannu, A.; Padua, A.; Mele, A.; Costa Gomes, M. Connecting chloride solvation with hydration in deep eutectic systems. *Phys. Chem. Chem.l Phys.* **2021**, *23*, 107–111. [CrossRef]
57. Chen, W.; Xue, Z.; Wang, J.; Zhao, X.; Mu, T. Investigation on the Thermal Stability of Deep Eutectic Solvents. *Acta Phys. Chim. Sin.* **2018**, *34*, 904–911. [CrossRef]
58. Delgado-Mellado, N.; Larriba, M.; Navarro, P.; Rigual, V.; Ayuso, M.; García, J.; Rodríguez, F. Thermal stability of choline chloride deep eutectic solvents by TGA/FTIR-ATR analysis. *J. Mol. Liq.* **2018**, *260*, 37–43. [CrossRef]
59. Škulcová, A.; Majová, V.; Dubaj, T.; Jablonský, M. Physical properties and thermal behavior of novel ternary green solvents. *J. Mol. Liq.* **2019**, *287*, 110991. [CrossRef]
60. Gontrani, L.; Plechkova, N.V.; Bonomo, M. In-Depth Physico-Chemical and Structural Investigation of a Dicarboxylic Acid/Choline Chloride Natural Deep Eutectic Solvent (NADES): A Spotlight on the Importance of a Rigorous Preparation Procedure. *ACS Sustain.Chem. Eng.* **2019**, *7*, 12536–12543. [CrossRef]
61. Rodriguez Rodriguez, N.; van den Bruinhorst, A.; Kollau, L.J.B.M.; Kroon, M.C.; Binnemans, K. Degradation of Deep-Eutectic Solvents Based on Choline Chloride and Carboxylic Acids. *ACS Sustain. Chem. Eng.* **2019**, *7*, 11521–11528. [CrossRef]
62. González-Rivera, J.; Husanu, E.; Mero, A.; Ferrari, C.; Duce, C.; Tinè, M.R.; D'Andrea, F.; Pomelli, C.S.; Guazzelli, L. Insights into microwave heating response and thermal decomposition behavior of deep eutectic solvents. *J. Mol. Liq.* **2020**, *300*, 112357. [CrossRef]
63. Cooper, E.R.; Andrews, C.D.; Wheatley, P.S.; Webb, P.B.; Wormald, P.; Morris, R.E. Ionic liquids and eutectic mixtures as solvent and template in synthesis of zeolite analogues. *Nature* **2004**, *430*, 1012–1016. [CrossRef]
64. Marcus, Y. Applications of Deep Eutectic Solvents. In *Deep Eutectic Solvents*; Marcus, Y., Ed.; Springer International Publishing: Cham, Switzerland, 2019; pp. 111–151.
65. Maschita, J.; Banerjee, T.; Savasci, G.; Haase, F.; Ochsenfeld, C.; Lotsch, B.V. Ionothermal Synthesis of Imide-Linked Covalent Organic Frameworks. *Angew. Chem. Int. Ed.* **2020**, *59*, 15750–15758. [CrossRef]
66. Gao, Z.; Xie, S.; Zhang, B.; Qiu, X.; Chen, F. Ultrathin Mg-Al layered double hydroxide prepared by ionothermal synthesis in a deep eutectic solvent for highly effective boron removal. *Chem. Eng. J.* **2017**, *319*, 108–118. [CrossRef]
67. Zhao, X.; Duan, W.; Wang, Q.; Ji, D.; Zhao, Y.; Li, G. Microwave-assisted ionothermal synthesis of Fe-LEV molecular sieve with high iron content in low-dosage of eutectic mixture. *Microporous Mesoporous Mater.* **2019**, *275*, 253–262. [CrossRef]
68. Wu, J.; Wang, Y.; Zhang, Y.; Meng, H.; Xu, Y.; Han, Y.; Wang, Z.; Dong, Y.; Zhang, X. Highly safe and ionothermal synthesis of Ti_3C_2 MXene with expanded interlayer spacing for enhanced lithium storage. *J. Energy Chem.* **2020**, *47*, 203–209. [CrossRef]
69. Zhang, G.; Lin, L.; Li, G.; Zhang, Y.; Savateev, A.; Zafeiratos, S.; Wang, X.; Antonietti, M. Ionothermal Synthesis of Triazine–Heptazine-Based Copolymers with Apparent Quantum Yields of 60% at 420 nm for Solar Hydrogen Production from "Sea Water". *Angew. Chem. Int. Ed.* **2018**, *57*, 9372–9376. [CrossRef]
70. Feng, Y.; Yan, G.; Wang, T.; Jia, W.; Zeng, X.; Sperry, J.; Sun, Y.; Tang, X.; Lei, T.; Lin, L. Synthesis of MCM-41-Supported Metal Catalysts in Deep Eutectic Solvent for the Conversion of Carbohydrates into 5-Hydroxymethylfurfural. *ChemSusChem* **2019**, *12*, 978–982. [CrossRef]
71. Chen, F.; Xie, S.; Huang, X.; Qiu, X. Ionothermal synthesis of Fe_3O_4 magnetic nanoparticles as efficient heterogeneous Fenton-like catalysts for degradation of organic pollutants with H_2O_2. *J. Hazard. Mater.* **2017**, *322*, 152–162. [CrossRef]
72. Hammond, O.S.; Edler, K.J.; Bowron, D.T.; Torrente-Murciano, L. Deep eutectic-solvothermal synthesis of nanostructured ceria. *Nat. Commun.* **2017**, *8*, 14150. [CrossRef] [PubMed]
73. Wu, J.; Lou, L.; Han, Y.; Xu, Y.; Zhang, X.; Wang, Z. Ionothermal synthesis of a photochromic inorganic–organic complex for colorimetric and portable UV index indication and UVB detection. *RSC Adv.* **2020**, *10*, 41720–41726. [CrossRef]
74. Xu, P.; Zheng, G.-W.; Zong, M.-H.; Li, N.; Lou, W.-Y. Recent progress on deep eutectic solvents in biocatalysis. *Biores. Bioproc.* **2017**, *4*, 34. [CrossRef]
75. Gao, F.; Huang, L.; Ma, Y.; Jiao, S.; Jiang, Y.; Bi, Y. Ionothermal synthesis, characterization of a new layered gallium phosphate with an unusual heptamer SBU. *J. Solid State Chem.* **2017**, *254*, 155–159. [CrossRef]
76. Zhao, H.; Xu, J.; Sheng, Q.; Zheng, J.; Cao, W.; Yue, T. $NiCo_2O_4$ Nanorods Decorated MoS_2 Nanosheets Synthesized from Deep Eutectic Solvents and Their Application for Electrochemical Sensing of Glucose in Red Wine and Honey. *J. Electrochem. Soc.* **2019**, *166*, H404–H411. [CrossRef]
77. Lin, Z.S.; Huang, Y. Tetraalkylammonium salt/alcohol mixtures as deep eutectic solvents for syntheses of high-silica zeolites. *Microporous Mesoporous Mater.* **2016**, *224*, 75–83. [CrossRef]
78. Lin, Z.S.; Huang, Y. Syntheses of high-silica zeolites in urea/choline chloride deep eutectic solvent. *Can. J. Chem.* **2016**, *94*, 533–540. [CrossRef]
79. Zheng, J.; Yu, Y.; Mo, Z.; Zhang, Z.; Wang, X.; Yin, S.; Peng, K.; Yang, Y.; Feng, X.; Cai, H. Industrial sector-based volatile organic compound (VOC) source profiles measured in manufacturing facilities in the Pearl River Delta, China. *Sci. Total Environ.* **2013**, *456–457*, 127–136. [CrossRef]
80. Montero-Montoya, R.; López-Vargas, R.; Arellano-Aguilar, O. Volatile Organic Compounds in Air: Sources, Distribution, Exposure and Associated Illnesses in Children. *Ann. Glob. Health* **2018**, *84*, 14. [CrossRef] [PubMed]

81. Vo, T.-D.-H.; Lin, C.; Weng, C.-E.; Yuan, C.-S.; Lee, C.-W.; Hung, C.-H.; Bui, X.-T.; Lo, K.-C.; Lin, J.-X. Vertical stratification of volatile organic compounds and their photochemical product formation potential in an industrial urban area. *J. Environ. Manag.* **2018**, *217*, 327–336. [CrossRef]
82. Villanueva, F.; Tapia, A.; Lara, S.; Amo-Salas, M. Indoor and outdoor air concentrations of volatile organic compounds and NO_2 in schools of urban, industrial and rural areas in Central-Southern Spain. *Sci. Total Environ.* **2018**, *622–623*, 222–235. [CrossRef]
83. Settimo, G.; Manigrasso, M.; Avino, P. Indoor Air Quality: A Focus on the European Legislation and State-of-the-Art Research in Italy. *Atmosphere* **2020**, *11*, 370. [CrossRef]
84. Heymes, F.; Manno-Demoustier, P.; Charbit, F.; Fanlo, J.L.; Moulin, P. A new efficient absorption liquid to treat exhaust air loaded with toluene. *Chem. Eng. J.* **2006**, *115*, 225–231. [CrossRef]
85. Darracq, G.; Couvert, A.; Couriol, C.; Amrane, A.; Thomas, D.; Dumont, E.; Andres, Y.; Le Cloirec, P. Silicone oil: An effective absorbent for the removal of hydrophobic volatile organic compounds. *J. Chem. Technol. Biotechnol.* **2010**, *85*, 309–313. [CrossRef]
86. Salar-García, M.J.; Ortiz-Martínez, V.M.; Hernández-Fernández, F.J.; de los Ríos, A.P.; Quesada-Medina, J. Ionic liquid technology to recover volatile organic compounds (VOCs). *J. Hazard. Mater.* **2017**, *321*, 484–499. [CrossRef] [PubMed]
87. Kudłak, B.; Owczarek, K.; Namieśnik, J. Selected issues related to the toxicity of ionic liquids and deep eutectic solvents—A review. *Environ. Sci. Pollut. Res.* **2015**, *22*, 11975–11992. [CrossRef]
88. Li, Z.; Wang, L.; Li, C.; Cui, Y.; Li, S.; Yang, G.; Shen, Y. Absorption of Carbon Dioxide Using Ethanolamine-Based Deep Eutectic Solvents. *ACS Sustain. Chem. Eng.* **2019**, *7*, 10403–10414. [CrossRef]
89. Mirza, N.R.; Nicholas, N.J.; Wu, Y.; Mumford, K.A.; Kentish, S.E.; Stevens, G.W. Experiments and Thermodynamic Modeling of the Solubility of Carbon Dioxide in Three Different Deep Eutectic Solvents (DESs). *J. Chem. Eng. Data* **2015**, *60*, 3246–3252. [CrossRef]
90. Leron, R.B.; Caparanga, A.; Li, M.-H. Carbon dioxide solubility in a deep eutectic solvent based on choline chloride and urea at T=303.15–343.15K and moderate pressures. *J. Taiwan Inst. Chem. Eng.* **2013**, *44*, 879–885. [CrossRef]
91. Lu, M.; Han, G.; Jiang, Y.; Zhang, X.; Deng, D.; Ai, N. Solubilities of carbon dioxide in the eutectic mixture of levulinic acid (or furfuryl alcohol) and choline chloride. *J. Chem. Thermodyn.* **2015**, *88*, 72–77. [CrossRef]
92. Moura, L.; Moufawad, T.; Ferreira, M.; Bricout, H.; Tilloy, S.; Monflier, E.; Costa Gomes, M.F.; Landy, D.; Fourmentin, S. Deep eutectic solvents as green absorbents of volatile organic pollutants. *Environ. Chem. Lett.* **2017**, *15*, 747–753. [CrossRef]
93. Yang, D.; Hou, M.; Ning, H.; Zhang, J.; Ma, J.; Yang, G.; Han, B. Efficient SO_2 absorption by renewable choline chloride–glycerol deep eutectic solvents. *Green Chem.* **2013**, *15*, 2261–2265. [CrossRef]
94. Deng, D.; Han, G.; Jiang, Y. Investigation of a deep eutectic solvent formed by levulinic acid with quaternary ammonium salt as an efficient SO_2 absorbent. *New J. Chem.* **2015**, *39*, 8158–8164. [CrossRef]
95. Liu, X.; Gao, B.; Deng, D. SO_2 absorption/desorption performance of renewable phenol-based deep eutectic solvents. *Sep. Sci. Technol.* **2018**, *53*, 2150–2158. [CrossRef]
96. Liu, B.; Wei, F.; Zhao, J.; Wang, Y. Characterization of amide–thiocyanates eutectic ionic liquids and their application in SO_2 absorption. *RSC Adv.* **2013**, *3*, 2470–2476. [CrossRef]
97. Zhang, K.; Ren, S.; Hou, Y.; Wu, W. Efficient absorption of SO_2 with low-partial pressures by environmentally benign functional deep eutectic solvents. *J. Hazard. Mater.* **2017**, *324*, 457–463. [CrossRef] [PubMed]
98. Akhmetshina, A.I.; Petukhov, A.N.; Mechergui, A.; Vorotyntsev, A.V.; Nyuchev, A.V.; Moskvichev, A.A.; Vorotyntsev, I.V. Evaluation of Methanesulfonate-Based Deep Eutectic Solvent for Ammonia Sorption. *J. Chem. Eng. Data* **2018**, *63*, 1896–1904. [CrossRef]
99. Zhong, F.-Y.; Peng, H.-L.; Tao, D.-J.; Wu, P.-K.; Fan, J.-P.; Huang, K. Phenol-Based Ternary Deep Eutectic Solvents for Highly Efficient and Reversible Absorption of NH_3. *ACS Sustain. Chem. Eng.* **2019**, *7*, 3258–3266. [CrossRef]
100. Zhong, F.-Y.; Huang, K.; Peng, H.-L. Solubilities of ammonia in choline chloride plus urea at (298.2–353.2)K and (0–300)kPa. *J. Chem. Thermodyn.* **2019**, *129*, 5–11. [CrossRef]
101. Li, Y.; Ali, M.C.; Yang, Q.; Zhang, Z.; Bao, Z.; Su, B.; Xing, H.; Ren, Q. Hybrid Deep Eutectic Solvents with Flexible Hydrogen-Bonded Supramolecular Networks for Highly Efficient Uptake of NH_3. *ChemSusChem* **2017**, *10*, 3368–3377. [CrossRef]
102. Deng, D.; Gao, B.; Zhang, C.; Duan, X.; Cui, Y.; Ning, J. Investigation of protic NH_4SCN-based deep eutectic solvents as highly efficient and reversible NH_3 absorbents. *Chem. Eng. J.* **2019**, *358*, 936–943. [CrossRef]
103. Di Pietro, M.E.; Colombo Dugoni, G.; Ferro, M.; Mannu, A.; Castiglione, F.; Costa Gomes, M.; Fourmentin, S.; Mele, A. Do Cyclodextrins Encapsulate Volatiles in Deep Eutectic Systems? *ACS Sustain. Chem. Eng.* **2019**, *7*, 17397–17405. [CrossRef]
104. Sun, S.; Niu, Y.; Xu, Q.; Sun, Z.; Wei, X. Efficient SO_2 Absorptions by Four Kinds of Deep Eutectic Solvents Based on Choline Chloride. *Ind. Eng. Chem. Res.* **2015**, *54*, 8019–8024. [CrossRef]
105. Kalepu, S.; Nekkanti, V. Insoluble drug delivery strategies: Review of recent advances and business prospects. *Acta Pharm. Sin. B* **2015**, *5*, 442–453. [CrossRef] [PubMed]
106. Savjani, K.T.; Gajjar, A.K.; Savjani, J.K. Drug Solubility: Importance and Enhancement Techniques. *ISRN Pharm.* **2012**, *2012*, 195727. [CrossRef] [PubMed]
107. Gala, U.; Pham, H.; Chauhan, H. Pharmaceutical Applications of Eutectic Mixtures. *J. Dev. Drugs* **2013**, *2*, 2.
108. Cherukuvada, S.; Nangia, A. Eutectics as improved pharmaceutical materials: Design, properties and characterization. *Chem. Commun.* **2014**, *50*, 906–923. [CrossRef]

109. Zainal-Abidin, M.H.; Hayyan, M.; Ngoh, G.C.; Wong, W.F.; Looi, C.Y. Emerging frontiers of deep eutectic solvents in drug discovery and drug delivery systems. *J. Control. Release* **2019**, *316*, 168–195. [CrossRef]
110. Pedro, S.N.; Freire, M.G.; Freire, C.S.R.; Silvestre, A.J.D. Deep eutectic solvents comprising active pharmaceutical ingredients in the development of drug delivery systems. *Expert Opin. Drug Deliv.* **2019**, *16*, 497–506. [CrossRef]
111. Rahman, M.S.; Roy, R.; Jadhav, B.; Hossain, M.N.; Halim, M.A.; Raynie, D.E. Formulation, structure, and applications of therapeutic and amino acid-based deep eutectic solvents: An overview. *J. Mol. Liq.* **2021**, *321*, 114745. [CrossRef]
112. Lu, C.; Cao, J.; Wang, N.; Su, E. Significantly improving the solubility of non-steroidal anti-inflammatory drugs in deep eutectic solvents for potential non-aqueous liquid administration. *MedChemComm* **2016**, *7*, 955–959. [CrossRef]
113. Li, Z.; Lee, P.I. Investigation on drug solubility enhancement using deep eutectic solvents and their derivatives. *Int. J. Pharm.* **2016**, *505*, 283–288. [CrossRef] [PubMed]
114. Olivares, B.; Martínez, F.; Rivas, L.; Calderón, C.; Munita, J.M.; Campodonico, P.R. A Natural Deep Eutectic Solvent Formulated to Stabilize β-Lactam Antibiotics. *Sci. Rep.* **2018**, *8*, 14900. [CrossRef] [PubMed]
115. Sut, S.; Faggian, M.; Baldan, V.; Poloniato, G.; Castagliuolo, I.; Grabnar, I.; Perissutti, B.; Brun, P.; Maggi, F.; Voinovich, D.; et al. Natural Deep Eutectic Solvents (NADES) to Enhance Berberine Absorption: An In Vivo Pharmacokinetic Study. *Molecules* **2017**, *22*, 1921. [CrossRef] [PubMed]
116. Aroso, I.M.; Silva, J.C.; Mano, F.; Ferreira, A.S.D.; Dionísio, M.; Sá-Nogueira, I.; Barreiros, S.; Reis, R.L.; Paiva, A.; Duarte, A.R.C. Dissolution enhancement of active pharmaceutical ingredients by therapeutic deep eutectic systems. *Eur. J. Pharm. Biopharm.* **2016**, *98*, 57–66. [CrossRef] [PubMed]
117. Abbott, A.P.; Ahmed, E.I.; Prasad, K.; Qader, I.B.; Ryder, K.S. Liquid pharmaceuticals formulation by eutectic formation. *Fluid Phase Equilib.* **2017**, *448*, 2–8. [CrossRef]
118. Stott, P.W.; Williams, A.C.; Barry, B.W. Transdermal delivery from eutectic systems: Enhanced permeation of a model drug, ibuprofen. *J. Control. Release* **1998**, *50*, 297–308. [CrossRef]
119. Nyqvist-Mayer, A.A.; Brodin, A.F.; Frank, S.G. Drug Release Studies on an Oil–Water Emulsion Based on a Eutectic Mixture of Lidocaine and Prilocaine as the Dispersed Phase. *J. Pharm. Sci.* **1986**, *75*, 365–373. [CrossRef]
120. Park, C.-W.; Mansour, H.M.; Oh, T.-O.; Kim, J.-Y.; Ha, J.-M.; Lee, B.-J.; Chi, S.-C.; Rhee, Y.-S.; Park, E.-S. Phase behavior of itraconazole–phenol mixtures and its pharmaceutical applications. *Int. J. Pharm.* **2012**, *436*, 652–658. [CrossRef]
121. Nazzal, S.; Smalyukh, I.I.; Lavrentovich, O.D.; Khan, M.A. Preparation and in vitro characterization of a eutectic based semisolid self-nanoemulsified drug delivery system (SNEDDS) of ubiquinone: Mechanism and progress of emulsion formation. *Int. J. Pharm.* **2002**, *235*, 247–265. [CrossRef]
122. Shen, Q.; Li, X.; Li, W.; Zhao, X. Enhanced Intestinal Absorption of Daidzein by Borneol/Menthol Eutectic Mixture and Microemulsion. *AAPS PharmSciTech* **2011**, *12*, 1044–1049. [CrossRef]
123. Wang, W.; Cai, Y.; Liu, Y.; Zhao, Y.; Feng, J.; Liu, C. Microemulsions based on paeonol-menthol eutectic mixture for enhanced transdermal delivery: Formulation development and in vitro evaluation. *Artif. Cells Nanomed. Biotechnol.* **2017**, *45*, 1241–1246. [CrossRef] [PubMed]
124. Duarte, A.R.C.; Ferreira, A.S.D.; Barreiros, S.; Cabrita, E.; Reis, R.L.; Paiva, A. A comparison between pure active pharmaceutical ingredients and therapeutic deep eutectic solvents: Solubility and permeability studies. *Eur. J. Pharm. Biopharm.* **2017**, *114*, 296–304. [CrossRef] [PubMed]
125. Woolfson, A.D.; Malcolm, R.K.; Campbell, K.; Jones, D.S.; Russell, J.A. Rheological, mechanical and membrane penetration properties of novel dual drug systems for percutaneous delivery. *J. Control. Release* **2000**, *67*, 395–408. [CrossRef]
126. Fiala, S.; Brown, M.B.; Jones, S.A. An investigation into the influence of binary drug solutions upon diffusion and partition processes in model membranes. *J. Pharm. Pharmacol.* **2008**, *60*, 1615–1623. [CrossRef]
127. Wang, H.; Gurau, G.; Shamshina, J.; Cojocaru, O.A.; Janikowski, J.; MacFarlane, D.R.; Davis, J.H.; Rogers, R.D. Simultaneous membrane transport of two active pharmaceutical ingredients by charge assisted hydrogen bond complex formation. *Chem. Sci.* **2014**, *5*, 3449–3456. [CrossRef]
128. Mota-Morales, J.D.; Gutiérrez, M.C.; Ferrer, M.L.; Sanchez, I.C.; Elizalde-Peña, E.A.; Pojman, J.A.; Monte, F.D.; Luna-Bárcenas, G. Deep eutectic solvents as both active fillers and monomers for frontal polymerization. *J. Polym. Sci. Part A Polym. Chem.* **2013**, *51*, 1767–1773. [CrossRef]
129. Sánchez-Leija, R.J.; Pojman, J.A.; Luna-Bárcenas, G.; Mota-Morales, J.D. Controlled release of lidocaine hydrochloride from polymerized drug-based deep-eutectic solvents. *J. Mater. Chem. B* **2014**, *2*, 7495–7501. [CrossRef]
130. Pradeepkumar, P.; Rajendran, N.K.; Alarfaj, A.A.; Munusamy, M.A.; Rajan, M. Deep Eutectic Solvent-Mediated FA-g-β-Alanine-co-PCL Drug Carrier for Sustainable and Site-Specific Drug Delivery. *ACS Appl. Bio. Mater.* **2018**, *1*, 2094–2109. [CrossRef]
131. Pradeepkumar, P.; Sangeetha, R.; Gunaseelan, S.; Varalakshmi, P.; Chuturgoon, A.A.; Rajan, M. Folic Acid Conjugated Polyglutamic Acid Drug Vehicle Synthesis through Deep Eutectic Solvent for Targeted Release of Paclitaxel. *ChemistrySelect* **2019**, *4*, 10225–10235. [CrossRef]
132. Aroso, I.M.; Craveiro, R.; Rocha, Â.; Dionísio, M.; Barreiros, S.; Reis, R.L.; Paiva, A.; Duarte, A.R.C. Design of controlled release systems for THEDES—Therapeutic deep eutectic solvents, using supercritical fluid technology. *Int. J. Pharm.* **2015**, *492*, 73–79. [CrossRef]
133. Silva, J.M.; Reis, R.L.; Paiva, A.; Duarte, A.R.C. Design of Functional Therapeutic Deep Eutectic Solvents Based on Choline Chloride and Ascorbic Acid. *ACS Sustain. Chem. Eng.* **2018**, *6*, 10355–10363. [CrossRef]

134. Chun, M.-K.; Hossain, K.; Choi, S.-H.; Ban, S.-J.; Moon, H.; Choi, H.-K. Development of cataplasmic transdermal drug delivery system containing eutectic mixture of lidocaine and prilocaine. *J. Pharm. Investig.* **2012**, *42*, 139–146. [CrossRef]
135. Serrano, M.C.; Gutiérrez, M.C.; Jiménez, R.; Ferrer, M.L.; del Monte, F. Synthesis of novel lidocaine-releasing poly(diol-co-citrate) elastomers by using deep eutectic solvents. *Chem. Commun.* **2012**, *48*, 579–581. [CrossRef] [PubMed]
136. Scherlund, M.; Brodin, A.; Malmsten, M. Nonionic Cellulose Ethers as Potential Drug Delivery Systems for Periodontal Anesthesia. *J. Colloid Interface Sci.* **2000**, *229*, 365–374. [CrossRef] [PubMed]
137. Mano, F.; Martins, M.; Sá-Nogueira, I.; Barreiros, S.; Borges, J.P.; Reis, R.L.; Duarte, A.R.C.; Paiva, A. Production of Electrospun Fast-Dissolving Drug Delivery Systems with Therapeutic Eutectic Systems Encapsulated in Gelatin. *AAPS PharmSciTech* **2017**, *18*, 2579–2585. [CrossRef]
138. Zakrewsky, M.; Banerjee, A.; Apte, S.; Kern, T.L.; Jones, M.R.; Sesto, R.E.D.; Koppisch, A.T.; Fox, D.T.; Mitragotri, S. Choline and Geranate Deep Eutectic Solvent as a Broad-Spectrum Antiseptic Agent for Preventive and Therapeutic Applications. *Adv. Healthc. Mater.* **2016**, *5*, 1282–1289. [CrossRef]
139. Mbous, Y.P.; Hayyan, M.; Wong, W.F.; Looi, C.Y.; Hashim, M.A. Unraveling the cytotoxicity and metabolic pathways of binary natural deep eutectic solvent systems. *Sci. Rep.* **2017**, *7*, 41257. [CrossRef]
140. García-Álvarez, J.; Hevia, E.; Capriati, V. The Future of Polar Organometallic Chemistry Written in Bio-Based Solvents and Water. *Chem. Eur. J.* **2018**, *24*, 14854–14863. [CrossRef]
141. Jablonský, M.; Škulcová, A.; Šima, J. Use of Deep Eutectic Solvents in Polymer Chemistry–A Review. *Molecules* **2019**, *24*, 3978. [CrossRef]
142. Grillo, G.; Calcio Gaudino, E.; Rosa, R.; Leonelli, C.; Timonina, A.; Grygiškis, S.; Tabasso, S.; Cravotto, G. Green Deep Eutectic Solvents for Microwave-Assisted Biomass Delignification and Valorisation. *Molecules* **2021**, *26*, 798. [CrossRef] [PubMed]
143. Chandran, D.; Khalid, M.; Walvekar, R.; Mubarak, N.M.; Dharaskar, S.; Wong, W.Y.; Gupta, T.C.S.M. Deep eutectic solvents for extraction-desulphurization: A review. *J. Mol. Liq.* **2019**, *275*, 312–322. [CrossRef]
144. Cai, T.; Qiu, H. Application of deep eutectic solvents in chromatography: A review. *TrAC Trends Anal. Chem.* **2019**, *120*, 115623. [CrossRef]
145. Vidal, C.; García-Álvarez, J.; Hernán-Gómez, A.; Kennedy, A.R.; Hevia, E. Introducing Deep Eutectic Solvents to Polar Organometallic Chemistry: Chemoselective Addition of Organolithium and Grignard Reagents to Ketones in Air. *Angew. Chem. Int. Ed.* **2014**, *53*, 5969–5973. [CrossRef]
146. Vidal, C.; García-Álvarez, J.; Hernán-Gómez, A.; Kennedy, A.R.; Hevia, E. Exploiting Deep Eutectic Solvents and Organolithium Reagent Partnerships: Chemoselective Ultrafast Addition to Imines and Quinolines Under Aerobic Ambient Temperature Conditions. *Angew. Chem. Int. Ed.* **2016**, *55*, 16145–16148. [CrossRef]
147. Sánchez-Condado, A.; Carriedo, G.A.; Presa Soto, A.; Rodríguez-Álvarez, M.J.; García-Álvarez, J.; Hevia, E. Organolithium-Initiated Polymerization of Olefins in Deep Eutectic Solvents under Aerobic Conditions. *ChemSusChem* **2019**, *12*, 3134–3143. [CrossRef] [PubMed]
148. Mallardo, V.; Rizzi, R.; Sassone, F.C.; Mansueto, R.; Perna, F.M.; Salomone, A.; Capriati, V. Regioselective desymmetrization of diaryltetrahydrofurans via directed ortho-lithiation: An unexpected help from green chemistry. *Chem. Commun.* **2014**, *50*, 8655–8658. [CrossRef]
149. Sassone, F.C.; Perna, F.M.; Salomone, A.; Florio, S.; Capriati, V. Unexpected lateral-lithiation-induced alkylative ring opening of tetrahydrofurans in deep eutectic solvents: Synthesis of functionalised primary alcohols. *Chem. Commun.* **2015**, *51*, 9459–9462. [CrossRef]
150. Ghinato, S.; Dilauro, G.; Perna, F.M.; Capriati, V.; Blangetti, M.; Prandi, C. Directed ortho-metalation–nucleophilic acyl substitution strategies in deep eutectic solvents: The organolithium base dictates the chemoselectivity. *Chem. Commun.* **2019**, *55*, 7741–7744. [CrossRef]
151. Arnodo, D.; Ghinato, S.; Nejrotti, S.; Blangetti, M.; Prandi, C. Lateral lithiation in deep eutectic solvents: Regioselective functionalization of substituted toluene derivatives. *Chem. Commun.* **2020**, *56*, 2391–2394. [CrossRef]
152. De Santi, V.; Cardellini, F.; Brinchi, L.; Germani, R. Novel Brønsted acidic deep eutectic solvent as reaction media for esterification of carboxylic acid with alcohols. *Tetrahedron Lett.* **2012**, *53*, 5151–5155. [CrossRef]
153. Nahar, Y.; Thickett, S.C. Greener, Faster, Stronger: The Benefits of Deep Eutectic Solvents in Polymer and Materials Science. *Polymers* **2021**, *13*, 447. [CrossRef]
154. Kalhor, P.; Ghandi, K. Deep Eutectic Solvents as Catalysts for Upgrading Biomass. *Catalysts* **2021**, *11*, 178. [CrossRef]
155. Qin, H.; Hu, X.; Wang, J.; Cheng, H.; Chen, L.; Qi, Z. Overview of acidic deep eutectic solvents on synthesis, properties and applications. *Green Energy Environ.* **2020**, *5*, 8–21. [CrossRef]
156. Habermas, K.L.; Denmark, S.E.; Jones, T.K. The Nazarov Cyclization. In *Organic Reactions*; John Wiley & Sons: Hoboken, NJ, USA, 1942; pp. 1–158.
157. Nejrotti, S.; Iannicelli, M.; Jamil, S.S.; Arnodo, D.; Blangetti, M.; Prandi, C. Natural deep eutectic solvents as an efficient and reusable active system for the Nazarov cyclization. *Green Chem.* **2020**, *22*, 110–117. [CrossRef]
158. Nejrotti, S.; Mannu, A.; Blangetti, M.; Baldino, S.; Fin, A.; Prandi, C. Optimization of Nazarov Cyclization of 2, 4-Dimethyl-1, 5-diphenylpenta-1, 4-dien-3-one in Deep Eutectic Solvents by a Design of Experiments Approach. *Molecules* **2020**, *25*, 5726. [CrossRef]

159. Contente, M.L.; Fiore, N.; Cannazza, P.; Roura Padrosa, D.; Molinari, F.; Gourlay, L.; Paradisi, F. Uncommon overoxidative catalytic activity in a new halo-tolerant alcohol dehydrogenase. *ChemCatChem* **2020**, *12*, 5679–5685. [CrossRef]
160. Sanyal, U.; Yuk, S.F.; Koh, K.; Lee, M.-S.; Stoerzinger, K.; Zhang, D.; Meyer, L.C.; Lopez-Ruiz, J.A.; Karkamkar, A.; Holladay, J.D.; et al. Hydrogen Bonding Enhances the Electrochemical Hydrogenation of Benzaldehyde in the Aqueous Phase. *Angew. Chem. Int. Ed.* **2021**, *60*, 290–296. [CrossRef]
161. Monna, T.; Fuhshuku, K.-I. Biocatalytic reductive desymmetrization of prochiral 1,3-diketone and its application to microbial hormone synthesis. *Mol. Catal.* **2020**, *497*, 111217. [CrossRef]
162. Yang, L.; Wurm, T.; Sharma Poudel, B.; Krische, M.J. Enantioselective Total Synthesis of Andrographolide and 14-Hydroxy-Colladonin: Carbonyl Reductive Coupling and trans-Decalin Formation by Hydrogen Transfer. *Angew. Chem. Int. Ed.* **2020**, *59*, 23169–23173. [CrossRef]
163. Hooshmand, S.E.; Afshari, R.; Ramón, D.J.; Varma, R.S. Deep eutectic solvents: Cutting-edge applications in cross-coupling reactions. *Green Chem.* **2020**, *22*, 3668–3692. [CrossRef]
164. Quivelli, A.F.; Vitale, P.; Perna, F.M.; Capriati, V. Reshaping Ullmann Amine Synthesis in Deep Eutectic Solvents: A Mild Approach for Cu-Catalyzed C–N Coupling Reactions With No Additional Ligands. *Front. Chem.* **2019**, *7*, 723. [CrossRef] [PubMed]
165. Vidal, C.; Suárez, F.J.; García-Álvarez, J. Deep eutectic solvents (DES) as green reaction media for the redox isomerization of allylic alcohols into carbonyl compounds catalyzed by the ruthenium complex [Ru(η3:η3-C10H16)Cl2(benzimidazole)]. *Catal. Commun.* **2014**, *44*, 76–79. [CrossRef]
166. Cicco, L.; Ríos-Lombardía, N.; Rodríguez-Álvarez, M.J.; Morís, F.; Perna, F.M.; Capriati, V.; García-Álvarez, J.; González-Sabín, J. Programming cascade reactions interfacing biocatalysis with transition-metal catalysis in Deep Eutectic Solvents as biorenewable reaction media. *Green Chem.* **2018**, *20*, 3468–3475. [CrossRef]
167. Ríos-Lombardía, N.; Rodríguez-Álvarez, M.J.; Morís, F.; Kourist, R.; Comino, N.; López-Gallego, F.; González-Sabín, J.; García-Álvarez, J. DESign of Sustainable One-Pot Chemoenzymatic Organic Transformations in Deep Eutectic Solvents for the Synthesis of 1,2-Disubstituted Aromatic Olefins. *Front. Chem.* **2020**, *8*, 139. [CrossRef] [PubMed]
168. Cavallo, M.; Arnodo, D.; Mannu, A.; Blangetti, M.; Prandi, C.; Baratta, W.; Baldino, S. Deep eutectic solvents as H2-sources for Ru(II)-catalyzed transfer hydrogenation of carbonyl compounds under mild conditions. *Tetrahedron* **2021**, *83*, 131997. [CrossRef]
169. Ríos-Lombardía, N.; Cicco, L.; Yamamoto, K.; Hernández-Fernández, J.A.; Morís, F.; Capriati, V.; García-Álvarez, J.; González-Sabín, J. Deep eutectic solvent-catalyzed Meyer–Schuster rearrangement of propargylic alcohols under mild and bench reaction conditions. *Chem. Commun.* **2020**, *56*, 15165–15168. [CrossRef] [PubMed]

Review

Ionic Liquids and Deep Eutectic Solvents for CO₂ Conversion Technologies—A Review

Kranthi Kumar Maniam [1] and Shiladitya Paul [1,2,*]

[1] Materials Innovation Centre, School of Engineering, University of Leicester, Leicester LE1 7RH, UK; km508@leicester.ac.uk
[2] Materials and Structural and Integrity Technology Group, TWI, Cambridge CB21 6AL, UK
* Correspondence: shiladitya.paul@twi.co.uk

Abstract: Ionic liquids (ILs) have a wide range of potential uses in renewable energy, including CO_2 capture and electrochemical conversion. With the goal of providing a critical overview of the progression, new challenges, and prospects of ILs for evolving green renewable energy processes, this review emphasizes the significance of ILs as electrolytes and reaction media in two primary areas of interest: CO_2 electroreduction and organic molecule electrosynthesis via CO_2 transformation. Herein, we briefly summarize the most recent advances in the field, as well as approaches based on the electrochemical conversion of CO_2 to industrially important compounds employing ILs as an electrolyte and/or reaction media. In addition, the review also discusses the advances made possible by deep eutectic solvents (DESs) in CO_2 electroreduction to CO. Finally, the critical techno-commercial issues connected with employing ILs and DESs as an electrolyte or ILs as reaction media are reviewed, along with a future perspective on the path to rapid industrialization.

Keywords: renewable energy; electrochemical conversion; carbon dioxide reduction; functionalized ionic liquids (ILs); carbon dioxide transformation

Citation: Maniam, K.K.; Paul, S. Ionic Liquids and Deep Eutectic Solvents for CO₂ Conversion Technologies—A Review. *Materials* **2021**, *14*, 4519. https://doi.org/10.3390/ma14164519

Academic Editors: Matteo Tiecco and Farooq Sher

Received: 29 May 2021
Accepted: 6 August 2021
Published: 11 August 2021

Publisher's Note: MDPI stays neutral with regard to jurisdictional claims in published maps and institutional affiliations.

Copyright: © 2021 by the authors. Licensee MDPI, Basel, Switzerland. This article is an open access article distributed under the terms and conditions of the Creative Commons Attribution (CC BY) license (https://creativecommons.org/licenses/by/4.0/).

1. Introduction

Controlling greenhouse gases (e.g., CO_2), which are often associated with energy generation and consumption, is the most challenging environmental issue and a source of great concern around the world [1,2]. Sustainable energy derived from renewable sources is an appealing option for mitigating global climate change. Significant efforts have been made to minimise reliance on fossil fuels by developing renewable energy sources but the contribution from these sources are in the range 25–28%, a 3% increase since 2019 [3,4]. As a renewable carbon source, converting or transforming CO_2 into an energy carrier as a fuel, fuel additive, or value-added chemical using renewable electricity could contribute to mitigating climate change and attaining a carbon-neutral economy [5,6]. Ionic liquids (ILs) are a new class of compounds that, due to their tunable physicochemical properties, have the potential to be used as novel materials in renewable energy applications such as CO_2 conversion [7–10].

During conversion, CO_2 is utilized as a feedstock to produce a wide range of fuels, fuel additives, acids, alcohols through formation of different chemical bonds (C–C, C–H, C–N, C–O). Direct reduction utilising homogeneous or heterogeneous catalysts converts CO_2 to CO and small organic molecules. Further, CO_2 is also converted to organic compounds such as carbonates, carboxylates, and carbamates by participating in an electrosynthesis process with an organic substrate (such as ethylene oxide, propylene oxide, olefins, and amines). Such synthesis is usually carried out in the presence of an alkylating agent, termed as CO_2 electro-organic transformation (referred as "transformation" in the review). Besides, storage of thermal energy in solar is one field where ILs due to their high temperature thermal stability are able to store a considerable amount of heat. Numerous ILs can theoretically be synthesised by combining different cations and anions, providing a good platform for design [11].

The performance of ILs can be credited to their ability to improve solubility, activation, and electrochemical conversion of CO_2 under moderate reaction conditions to fuels. This makes ILs appealing as an alternative media. Furthermore, typical electrolytes used for CO_2 conversion have drawbacks such as high volatility, separation behaviour, corrosivity to metals, and instability in electroreduction or transformation, which motivated the scientific community to focus on ILs [7,10–14]. There have recently been numerous studies on the utilisation of CO_2-saturated ILs as electrolytes to enhance the electroreduction of CO_2 and electrosynthesis of organic chemicals (such as carbonates, carbamates, etc). Recently, ILs have been explored as thermal energy storage fluids, and are becoming a promising research topic [11]. Because of their high ionic conductivities and wide electrochemical windows, as well as their increased solubility relative to traditional solvents, ILs can play a critical role in CO_2 conversion. Figure 1 depicts the use of ILs in renewable energy applications.

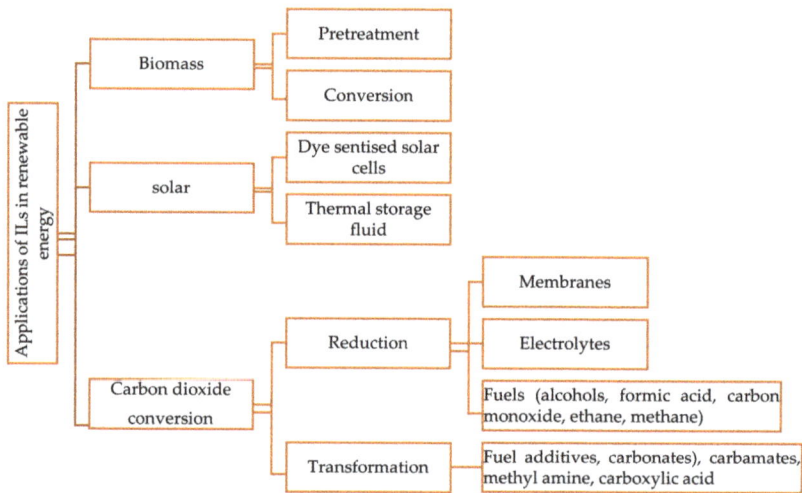

Figure 1. Applications of ionic liquids in renewable energy.

This article focuses on improvements in the role of ILs, DESs in renewable resources, such as CO_2 reduction and transformation to value added chemicals. The primary goal is to provide a critical review of the new challenges and prospects of ILs for adopting sustainable processes of renewable energy. CO_2 utilization via electrochemical route has been, and is a hot topic, with ILs, gaining increased attention. There are a lot of reviews published in the electrochemical reduction of CO_2 using ILs [10,15] and the electrosynthesis of organic compounds using CO_2 [8,9]. However, reviews on the combination of CO_2 reduction and transformation employing ILs and deep eutectic solvents (DESs) are very scarce. The current review presents a synopsis of the considerable progress achieved by ILs in renewable energy applications, with a focus on CO_2 conversion and transformation. In addition, the new developments associated with deep eutectic solvents (DESs) are also covered in this review. Unlike the other reviews in the field, we have taken a different approach in consolidating the results from the fields proposed. Additional information on the technical challenges, cost considerations that are expected to be solved, which were not sufficiently covered in previous reviews and works, will be presented. The ionic liquids considered in this review are listed in Table 1.

Table 1. List of ionic liquids (ILs) discussed in this review.

Ionic Liquid	Chemical Formula	Cation	Anion	Abbreviation	Ref.
1–butyl–3–methyl imidazolium bromide	$[(C_8H_{15}N_2Br)]$	$[(C_4H_9)(CH_3)(C_3H_3N_2)]^+$	$[Br]^-$	[BMIm][Br]	[16]
1–butyl–3–methyl imidazolium chloride	$[(C_8H_{15}N_2Cl)]$	$[(C_4H_9)(CH_3)(C_3H_3N_2)]^+$	$[Cl]^-$	[BMIm][Cl]	[17]
1–ethyl–3–methyl imidazolium bis(trifluoro methyl sulfonyl) imide	$[(C_8H_{11}F_6N_3O_4S_2)]$	$[(C_2H_5)(CH_3)(C_3N_2H_4H)]^+$	$[CF_3SO_2CF_3SO_2N]^-$	[EMIm][Tf$_2$N]	[18]
1–propyl–1–methyl imidazolium bis(trifluoro methyl sulfonyl) imide	$[(C_9H_{13}F_6N_3O_4S_2)]$	$[(C_3H_7)(CH_3)(C_3N_2H_4H)]^+$	$[CF_3SO_2CF_3SO_2N]^-$	[PMIm][Tf$_2$N]	[18]
1–butyl–3–methyl imidazolium bis(trifluoro methyl sulfonyl) imide	$[(C_{10}H_{15}F_6N_3O_4S_2)]$	$[(C_4H_9)(CH_3)(C_3N_2H_4H)]^+$	$[CF_3SO_2CF_3SO_2N]^-$	[BMIm][Tf$_2$N]	[19]
1–butyl–3–methyl imidazolium hexa fluorophosphate	$[(C_8H_{15}F_6N_2P)]$	$[(C_4H_9)(CH_3)(C_3N_2H_4H)]^+$	$[PF_6]^-$	[BMIm][PF$_6$]	[19]
1–butyl–3–methyl imidazolium tetrafluoroborate	$[(C_8H_{15}F_4N_2B)]$	$[(C_4H_9)(CH_3)(C_3N_2H_4H)]^+$	$[BF_4]^-$	[BMIm][BF$_4$]	[19]
1–butyl–3–methyl imidazolium trifluoro methane sulfonate	$[(C_9H_{15}F_3N_2O_3S)]$	$[(C_4H_9)(CH_3)(C_3N_2H_4H)]^+$	$[CF_3SO_3]^-$	[BMIm][TfO]	[19]
1–butyl–3–methylimidazolium tris(perfluoroethyl)trifluorophosphate	$C_{14}H_{15}FN_2P$	$[(C_4H_9)(CH_3)(C_3N_2H_4H)]^+$	$[(C_2F_5)_3PF_3]^-$	[BMIm][FAP]	[19]
1–butyl–3–methylimidazolium dicyanamide	$[(C_{10}H_{15}N_5)]$	$[(C_4H_9)(CH_3)(C_3N_2H_4H)]^+$	$[C_2N_3]^-$	[BMIm][DCA]	[19]
1–butyl–1–methyl pyrrolidine bis(trifluoro methyl sulfonyl) imide	$[(C_{11}H_{20}F_6N_2O_4S_2)]$	$[(C_4H_9)(CH_3)C_4H_9NH)]^+$	$[CF_3SO_2CF_3SO_2N]^-$	[BMPyl][Tf$_2$N]	[20]
1–ethyl–3–methyl imidazolium hexa fluorophosphate	$[(C_6H_{11}F_6N_2P)]$	$[(C_2H_5)(CH_3)(C_3N_2H_4H)]^+$	$[PF_6]^-$	[EMIm][PF$_6$]	[21]
1–ethyl–3–methyl imidazolium tetrafluoroborate	$[(C_6H_{11}F_4N_2B)]$	$[(C_2H_5)(CH_3)(C_3N_2H_4H)]^+$	$[BF_4]^-$	[EMIm][BF$_4$]	[21]
1–ethyl–3–methyl imidazolium trifluoro methane sulfonate	$[(C_7H_{11}F_3N_2O_3S)]$	$[(C_2H_5)(CH_3)(C_3N_2H_4H)]^+$	$[CF_3SO_3]^-$	[EMIm][TfO]	[21]
1–butyl–3–methyl imidazolium acetate	$[(C_{10}H_{18}N_2O_2)]$	$[(C_4H_9)(CH_3)(C_3N_2H_4H)]^+$	$[CH_3COO]^-$	[BMIm][Ac]	[22]

2. Carbon Dioxide Conversion in ILs

2.1. Conversion by Electroreduction

Although the use of CO_2 as a raw material appears to be particularly promising, CO_2's inert nature, along with its high thermodynamic stability, poses a challenge to CO_2 conversion, transformation, and utilisation as an effective renewable energy source. Numerous CO_2 reduction strategies, such as thermal, biochemical, photochemical, and electrochemical approaches, have been studied extensively, with various degrees of success and practicality [23]. Among them, electrochemical CO_2 conversion, either by electroreduction or electro-transformation, to value-added chemicals and fuels has sparked great interest for sustainable energy conversion and storage [24]. The key advantages with the electrochemical conversion of CO_2 is the tunability of the reactions by adjusting the electrolytes, operating conditions and electrode materials.

As an important component in the electroreduction process, the electrolyte interacts with the electrode surface, reactants, and intermediates, which plays a critical role in charge transport [25]. Depending on the electrochemical reduction routes involved in the process, changes in CO_2 solubility, conductivity, and viscosity are thought to have a substantial impact on the catalytic activity for electroreduction and electro-transformation into alcohols, alkanes, alkenes, acetates, formates, and organic carbonates (cyclic, dialkyl). Figure 2 depicts the possible formation of CO_2 electroreduction products and their accompanying redox potentials (vs. SHE at pH = 7) along with the applications of the relevant products. While the primary role of an electrolyte is to conduct the ionic charge between the electrodes, it needs to satisfy certain criteria such as (i) solubility for CO_2, (ii) compatibility with electrode materials (especially cathodes), (iii) stability (without decomposition), (iv) safety in handling and storage. Achieving these criteria will not only make them function as a good electrolyte but also improve the overall efficiency of the process. Because of their high ionic conductance, aqueous electrolytes containing inorganic salts are the most often employed electrolytes. However, they have poor CO_2 solubility (0.03 mol L^{-1} CO_2 in water under 298 K, 0.1 MPa), low conversion rates (30 percent at 1 A cm^{-2}) with substantial hydrogen evolution as a side reaction, and an unsatisfactory applicable potential range [26].

Carbon dioxide reduction	Reaction	Applications
	$CO_2 + 2H^+ + 2e^- \rightarrow CO + H_2O; E_{redox} = -0.53$ V $2H^+ + 2e^- \rightarrow H_2; E_{redox} = -0.41$ V	Fuel production, metal refining, electronic, semiconductor industry
	$CO_2 + 2H^+ + 2e^- \rightarrow HCOOH; E_{redox} = -0.61$ V	leather tanning, fuel cells, dyeing, textiles, preservatives, intermediate
	$CO_2 + 4H^+ + 4e^- \rightarrow HCHO + H_2O; E_{redox} = -0.48$ V	industrial resins, adhesives (plywood, carpeting), particle board, coatings, textiles, precursor, fuel cells
	$CO_2 + 6H^+ + 6e^- \rightarrow CH_3OH + H_2O; E_{redox} = -0.38$ V	Fuel cells, energy carrier, polymers, olefins, amines, gasoline, fuel
	$CO_2 + 8H^+ + 8e^- \rightarrow CH_4 + 2H_2O; E_{redox} = -0.24$ V	rocket fuel, electricity generation, steam reforming
	$CO_2 + 12H^+ + 12e^- \rightarrow C_2H_5OH + 3OH^-; E_{redox} = -0.33$ V	engine fuel, antiseptic, antidote, fuel cells, household heating
	$2CO_2 + 12H^+ + 12e^- \rightarrow C_2H_4 + 4H_2O; E_{redox} = -0.34$ V	Food, chemicals, glass, metal fabrication, refining, rubber plastics

Figure 2. Summary of the possible CO_2 reductions with their redox potentials and applications.

2.1.1. CO_2 Electroreduction in Ionic Liquids

To overcome the limitations of aqueous electrolytes, such as poor CO_2 solubility, hydrogen evolution at the cathode, and low conversion efficiency, organic solvents such as acetonitrile, dimethyl sulfoxide, polycarbonates, and dimethylformide were utilised as non-aqueous systems to improve CO_2 conversion. These have been shown to offer better CO_2 solubility than aqueous electrolytes while also having good recyclability. However, significant shortcomings of organic solvent electrolytes, such as their high volatility, and poor safety characteristics (flammability, toxicity), have limited their commercialization and applicability. Furthermore, the recycling costs of organic solvents employed in electrolyzers remain high due to their potential miscibility with the target products [27]. This motivated the scientific community to pursue research in ionic liquids as alternate potential electrolytes to organic solvents, aqueous media (containing inorganic salts). The key benefits are their tunable features, such as polarity, hydrophobicity, and solvent miscibility, which can be achieved by modifying the arrangement of the cations and anions [7]. Ionic liquids have recently attracted a lot of interest as electrolytes because of their high CO_2 adsorption capacity, solubility, selectivity to CO_2 over other gases (such as N_2, O_2, CH_4), and low energy consumption. They are a very diverse and efficient family of promoters in the electrochemical reduction of CO_2, with the ability to improve reaction characteristics by changing their interaction and exhibiting high electrochemically stable windows (4–5 V), high ionic conductivities, and low vapour pressures [10,15]. Also, it has been observed that ILs can lower the energy barrier of reactions by building complexes with the intermediates generated during the reaction [28].

Zhao et al. [29] employed a [BMIm][PF$_6$] electrolyte to make syngas (CO + H$_2$), a value-added fuel, and demonstrated the promise of an ionic liquid for CO$_2$ conversion applications. This was claimed to be one of the earliest works that employed ionic liquid as an electrolyte. Besides, the study confirmed the synthesis of small organic compounds such as formic acid in lower concentrations. Rosen et al. used an IL-mediated selective conversion strategy to reduce the high overpotential that is commonly observed with CO$_2$ reduction to CO in aqueous systems, where the IL was demonstrated to lower the energy required to generate the (CO$_2$•$^-$) intermediate [28]. When [EMIm][BF$_4$] was introduced as an electrolyte to the reaction system, electrochemical characterization results showed that the overpotential could be reduced by up to 0.2 volts with silver serving as the cathode. The authors attributed the reduction in overpotential associated with the electroreduction of CO$_2$ to CO to the complexation between CO$_2$ and [BF$_4$]$^-$. This complex was shown to play a critical role in lowering the energy that is required to break the chemical bonds in CO$_2$ to form the (CO$_2$•$^-$) intermediate, and achieved a continuous production of CO up to 7 hours with a Faradaic efficiency of ~96% [30]. Since then attempts were made with imidazolium cations with different anions [BF$_4$]$^-$, [CH$_3$COO]$^-$, [Tf$_2$N]$^-$, [PF$_6$]$^-$, [TfO]$^-$ using different catalyst systems such as 2D dichalcogenide structures (MoS$_2$ [31], WS$_2$ [32]), doped carbons [33], metals, metal-alloys. One of the significant works was reported by Sun et al. [19] which focused on imidazolium based ILs with different anions: [BF$_4$]$^-$, [PF$_6$]$^-$, [TfO]$^-$, [Tf$_2$N]$^-$,and [DCA]$^-$ with [BMIm]$^+$ as the common cation. N-doped carbon on carbon paper which can exhibit the feature of a graphene was used the catalyst. The results demonstrated a conversion of CO$_2$ into CH$_4$ with fluorine-based ones displaying higher total current densities than the non-fluorine ones. The authors explained this to the strong interactions between CO$_2$ and fluorine, which weakens the C=O bond by forming a Lewis acid-base adduct and the fact was also supported by other reference works. In addition to the typical ILs with "common" anions, Snuffin et al. developed and synthesized a novel imidazolium based IL with dual halide anion combination: 1–ethyl–3–methyl–imidazolium trifluorochloroborate [EMIm][BF$_3$Cl], demonstrated a strong CO$_2$ solubility and also a positive reduction potential of −1.8 V while promoting electroreduction of CO$_2$ [34].

Although there are many advantages in using ILs as electrolytes in electroreduction of CO$_2$ besides excellent physico-chemical properties such as high reactant solubility, lowering the energy barrier, the relatively high cost and their associated viscosity of ILs hinder their practical application. Table 2 provides a summary of some ILs reported as electrolytes for electroreduction of CO$_2$ with different catalyst combinations. As can be seen from the Table 2, it is clear that imidazolium-based ILs have been the most investigated type of IL. Especially, [EMIm][BF$_4$], followed by [BMIm][BF$_4$] and [BMIm][PF$_6$], have been, by far, the most widely used ILs in the electroreduction of CO$_2$. [BF$_4$]$^-$, [PF$_6$]$^-$, [TfO]$^-$, and [Tf$_2$N]$^-$ are the most commonly used anions as can be seen from the table and reported by many works. This can be related to the Lewis acid-base interaction effect between the selected anion X$^-$ (X: [BF$_4$]$^-$, [PF$_6$]$^-$, [TfO]$^-$, [Tf$_2$N]$^-$) and CO$_2$ molecule forming [X$^-$CO$_2$]$^-$ complex. Such a complex displays strong alkalinity and tends to displace the bonds that exist between the inert anions (B−F, P−F, C−F, and S=O). Also, these anions possess weak ionization characteristics and weak coordination capacity and as a result promote the electrochemical reduction of CO$_2$ by favouring the interactions between CO$_2$ and metal electrode without affecting the reaction characteristics. Amongst the anions, imidazolium cations with [BF$_4$]$^-$ anions were reported high Faradaic efficiencies of >98% owing to their strongest Lewis acid-base interactions.

Table 2. Summary of IL-based electrolytes used in the electroreduction of CO_2.

Electrolyte	Catalyst	Reactor Type	Major Products	Faradaic Efficiency (FE, %) [1]	Reference
[BMIm]BF$_4$	N-doped carbon (graphene-like) materials/carbon paper electrodes	H-Cell	CH_4	93.50	[19]
[BMIm][Ac]	Platinum	Two electrode cell	Oxalate, CO, carbonate	-	[22]
18 mol % [EMIm][BF$_4$] in H$_2$O	Silver nanoparticles	Flow cell	CO, H$_2$	96	[28]
[BMim][PF$_6$]	Copper plank	High pressure undivided cell	CO, H$_2$, HCOOH (traces)	90.20	[29]
4 mol % [EMIm][BF$_4$] + 96 mol % H$_2$O	Molybdenum disulphide	Custom made 2 compartment three electrode cell	CO	98	[31]
[EMIm][BF$_4$]/H$_2$O (50 vol %/50 vol %)	WSe$_2$ nanoflakes	2 compartment 3-electrode electrochemical cell	C/O	24	[32]
25 mol % [EMIm][BF$_4$] + H$_2$O (75 mol %)	Metal free carbon nanofibers	3-electrode electrochemical cell	CO	98	[33]
[EMIm][BF$_4$]: H$_2$O (1:1 v/v)	Nanostructured and nanosized Titania	H-cell	Low density poly ethylene (LDPE)	14	[35]
10.5 mol % [EMIm][BF$_4$] + 89.5 mol % H$_2$O	Silver nanoparticles	Flow cell	CO	100	[36]
[EMIm][BF$_4$]	Silver nanoparticles	Flow cell	CO	-	[37]
[BMIm][BF$_4$]	Flat platinum spirals	2-compartment homemade glass cell	NHC [2]–CO$_2$ adduct	-	[38]
80 wt % [BMIm][Cl] + 20 wt % H$_2$O	Silver	H-Cell	CO	>99	[39]
[BMIm][BF$_4$]	Indium tin oxide	Undivided glass electrochemical cell	CO	64.90	[40]
[EMIm][Tf$_2$N]	Pre-anodized Pt electrode	Two electrode cell	HCOOH	-	[41]
[BMPyr][Tf$_2$N]	Pre-anodized Pt electrode	Two electrode cell	HCOOH	-	[41]
[EMIm][BF$_4$]/H$_2$O (92/8 v/v %)	Silver nanoflowers	Flow cell	CO	75	[42]

[1] The values are indicated based on the optimized conditions reported by the referenced works. [2] NHC: N-heterocyclic carbene.

The Tanner group [18] investigated the effect of several cations: [EMIm]$^+$, [BMIm]$^+$, [PMIm]$^+$, [BMPyl]$^+$ on the performance of electrochemical CO_2 reduction using silver electrodes as the catalyst. Since comparison, analysis of the data in terms of potentials could not demonstrate the influence of imidazolium cations towards the performance, studies were extended with different anions: [BF$_4$]$^-$, [Tf$_2$N]$^-$, [FAP]$^-$ with [BMIm]$^+$ as the cation. [BMIm][FAP] based IL displayed the best reactant solubility amongst the others but with a lower current density. Bruzon et al. [43] investigated CO_2 electroreduction in nitrogen-based imidazolium-based ILs with [FAP]$^-$ as the anion, observed a significant reduction in the electric potential which is subsequently utilized to reduce CO_2. It has been demonstrated that the functional groups: –OCH$_3$, –CN minimised the free energy to form the first intermediate of CO_2 reduction, ($CO_2 \bullet ^-$) to a greater extent. Since the mechanism is not clearly understood, it is widely assumed that the structure of the IL might have more influence on the CO_2 reaction than the reactant solubility.

2.1.2. CO_2 Electroreduction in Deep Eutectic Solvents (DESs)

Deep eutectic solvents (DESs) are obtained by combining a hydrogen bond acceptor and donor in specific mole ratios. These mixtures exhibit low melting points, similar properties and characteristics to ILs that are required for the electrochemical reduction of CO_2 [44]. Also, these are less expensive and considered to be potential alternatives to ILs. Verma et al. [45] conducted experiments employing [EMIm]-based ILs and choline chloride:urea (ChCl:Urea—1:2) DESs as the electrolyte media for the electrochemical reduction of CO_2. The results showed low conductivity and performance which increased on adding potassium chloride (KCl) to the non-aqueous electrolyte. Vasilyev et al. [17] studied the electrochemical reduction of CO_2 employing different ChCl-based DESs, imidazolium chloride based DESs. Imidazolium chloride-based DESs were prepared by mixing IL chloride with ethylene glycol, polyethylene glycol-200 (PEG-200) as hydrogen bond donors. The results demonstrated that choline-based DESs, IL-chlorides with EG facilitated the electrochemical reduction of CO_2, when silver is used as the catalyst. However, certain mixtures were shown to be non-room temperature liquids, which on addition of organic solvents could facilitate the electrochemical reduction of carbon dioxide with improved reactant (CO_2) solubilities. Hydrogen evolution reaction was observed by Vasilyev et al. [17], Verma et al. [45] when water up to 15 vol % was added to ChCl-based DESs while displaying high FE with CO as the major product. The presence of the hydroxy group in the structure of the imidazolium cations, choline based DESs, was shown to be the key factor in enhancing the electrochemical reduction of CO_2.

In recent times, the modification or preparation of catalytic electrode materials using ILs/DESs have gained primary attention as they are expected to reduce the background current of the electrode, optimize the performance of the electrode materials and favour the catalytic reduction of CO_2. Besides, ILs/DESs can also be used as a medium to prepare catalysts. Bohlen et al. [46] performed the electrodeposition of indium from 1:2 M choline based DES (ChCl:EG—1:2), employed them as an electrocatalyst for the electrochemical reduction of CO_2 to formate. As per the Cui et al. [15] review, this was the first publication which reported on the preparation of CO_2 reduction catalysts by electrodeposition in DESs. Extending this method by tailoring the DES electrolyte properties and the electrodeposition conditions, it is possible to develop other metals with different sizes, shapes and structures and faces.

This provides a new research idea to produce selective products such as ethylene for future exploration in this field considering the combined advantages with ILs/DESs and electrodeposition. Table 3 lists few of the works studied using choline-based DESs and imidazolium chloride-based ones with different hydrogen bond donors. One common feature amongst all the studied DESs is the reactor type and the major product associated with the CO_2 electroreduction. Figure 3 highlights the developments of the ILs, DESs that are employed for the electroreduction of CO_2 using different classes of catalysts.

Table 3. Summary of DESs used in the electroreduction of CO_2.

Electrolyte	Catalyst	Reactor Type	Major Products	Faradaic Efficiency (FE, %)	Reference
[ChCl] [1] [Urea] (1:2)	Silver	U-type divided cell	CO	15.80	[17]
[ChCl][Urea] (1:2) + H_2O (15 vol %)	Silver	U-type divided cell	CO	59	[17]
[ChCl]–EG [1] (1:2)	Silver	U-type divided cell	CO	78	[17]
[BMIm][Cl]:[EG] (1:2)	Silver	U-type divided cell	CO	95.80	[17]
1M [ChCl] in EG	Silver	U-type divided cell	CO	71.10	[17]
1M [ChCl] in PEG-200	Silver	U-type divided cell	CO	83.20	[17]
1M [BMIm][Cl] in PEG-200	Silver	U-type divided cell	CO	85.90	[17]
2M [ChCl][Urea] (1:2)	Silver	Electrochemical flow reactor	CO	94.10	[45]
[ChCl][Urea] (1:2) + H_2O (50 vol %)	Silver	H-cell	CO	96	[47]
[MEAHCl][MDEA] [2]	Silver	Three electrode cell	CO	71	[48]

[1] ChCl: choline chloride; EG: ethylene glycol. [2] [MEAHCl][MDEA]: [monoethanolamine hydrochloride] [methyl diethanolamine].

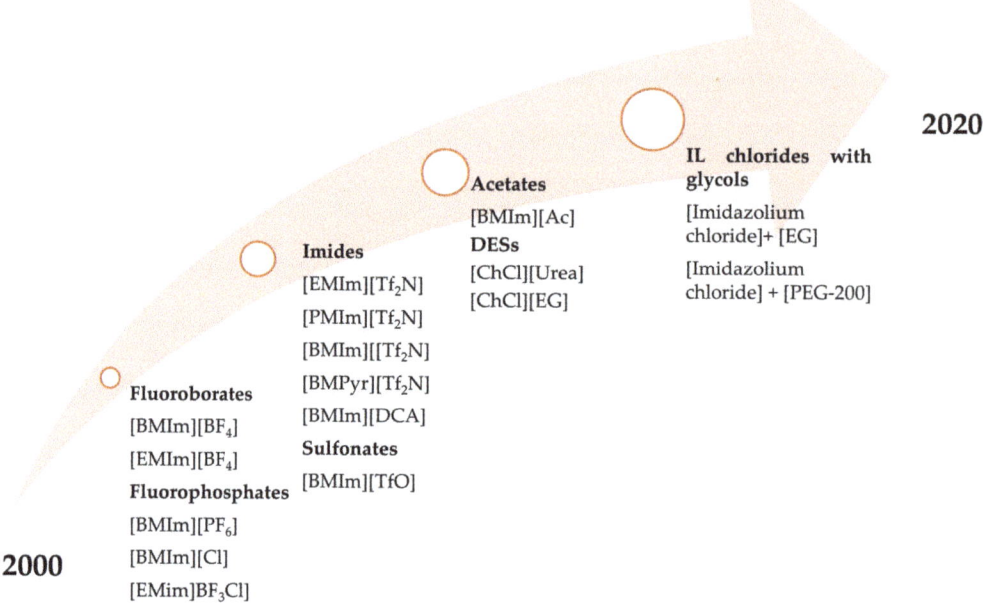

Figure 3. Developments of the ILs, DESs in CO_2 electroreduction covering the period 2000–2020.

2.2. Conversion by Electrotransformation

An another effective method of utilising CO_2 is to electrosynthesize the C_1 feedstock into valuable fuel additives without the use of a hydrogen source [9,49]. In general, this non-hydrogenation process with mild working conditions converts CO_2 into a diverse array of organic compounds such as carbonates (cyclic, dialkyl), carboxylic acids, and carbamates. Carbonates, specifically cyclic and dialkyl ones were identified to be the most effective fuel additives. It is possible to electrosynthesise numerous kinds of products via the transformation pathway through electrochemical processes involving CO_2. A wide variety of substrates such as epoxides, alcohols, amines, aryl halides, and olefins have been used to transform the CO_2 into their respective organic compounds [9,50–52]. This approach is thought to be the most efficient for some reactions that are thermodynamically unfavourable in the absence of external energy and when thermal catalysis options are severely constrained. The use of an electrochemical approach to synthesise organic compounds has several merits, including moderate conditions, high functional group tolerance, and inherent scalability and sustainability [8]. The general pathway of electrosynthesis of products (such as carbonates, carbamates, and carboxylates) via CO_2 transformation involves the generation of electro-induced radical/anion from CO_2-saturated with ILs and/or substrates. Subsequently, the generated radical/anion reacts with other substrates, yielding either of the transformed products mentioned above. Several investigations on the studies performed in this field have further shown that ILs have a stabilization effect on the electro-induced CO_2 molecule or substrates (epoxides, ketones, alkenes, etc) radical/anion. When combined with the high reactant CO_2 solubility and favourable electrochemical properties required for CO_2 transformation, ILs are viewed as an alternative eco-friendly and prospective reaction medium to the currently utilized hazardous volatile organic solvents [51]. Previous review papers [8,9,50,52,53] address the significant research published in this domain, with a focus on the electrosynthesis of carbonates via the transformation route. As a result, this section presents a succinct overview of the most recent developments in the value-added chemical compounds generated by electrosynthesis via the CO_2 transformation route in ILs other than carbonates. Figure 4 depicts the product classes that are generally produced from electrosynthesis via the transformation pathway utilising CO_2-saturated with ILs.

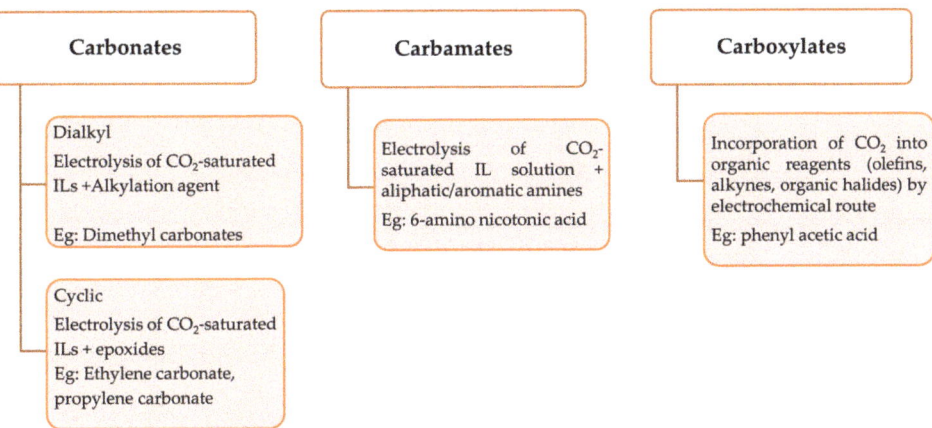

Figure 4. Summary of the organic products that were obtained from electrosynthesis via the CO_2 transformation route [50].

Carbonates produced from electrosysnthesis reactions are obtained by the transformation of CO_2 with organic substrates (epoxides, olefins). These provide a green synthesis pathway via the C—O bond formation which avoid the use of phosgene or CO, that are generally toxic in nature. These can be cyclic or dialkyl and have gained widespread interest in applications such as battery electrolytes, intermediates, polar aprotic solvents, and so on. Typical examples of cyclic carbonates include propylene carbonate, methylene carbonate while dimethyl carbonate stands out to be a classic example of a diakyl carbonate. Yuan group performed the electrosynthesis of cyclic carbonates utilising ILs saturated with CO_2 and epoxide [54]. Under mild operating conditions, the reaction was studied using an undivided cell with a Cu working electrode and an Al or Mg rod sacrificial anode. The performance of CO_2 cycloaddition to various epoxide substrates (propylene oxide, epichlorohydrin, and styrene oxide) was evaluated.

The best results were shown to be obtained by using propylene oxide as the substrate and [BMIm][BF$_4$] as the reaction media, which resulted in 92% conversion and 100% selectivity to the desired product (cyclopropylene carbonate). CO_2 underwent a one-electron reduction to generate the ($CO_2 \bullet^-$) radical anion, which subsequently interacted with the activated substrate to produce the matching cyclic carbonate. Wang et al. reported the electrosynthesis of butylene carbonate (cyclic carbonate) from diols (1,2-butanediol) in CO_2-saturated imidazolium based IL ([BMIm][BF$_4$]) in an undivided cell under mild operating conditions (1 atm, 50 °C). However, the highest yield that could be achieved employing the proposed IL system was 12%, with magnesium anode and copper as the cathode [55]. Under mild conditions, at atmospheric pressure and temperatures of 55 °C, Zhang et al. [56] developed a novel electrochemical technique to synthesise dialkyl carbonates from a CO_2-saturated imidazolium based IL:[BMIm][BF$_4$] in the presence of an alkylating agent. The corresponding process eliminates the use of organic solvents and supporting electrolytes. Most of the literature demonstrated that the CO_2 transformation in IL was shown to be relatively easier than that in organic solvents. This was attributed to the formation of a ([BMIm]$^+$–[CO_2^-]) ion-pair [56], which could facilitate the better yield of the carbonates from electrosynthesis.

Dimethyl carbonates are considered to be an important class of dialkyl carbonates which serve as methylation agent and replace toxic substances such as phosgene, tertiary butyl methyl ether, methyl halides. The Wu group [55] demonstrated the utilization of N-heterocyclic carbenes (NHC), electrogenerated from CO_2-saturated [BMIm][X]—ILs ([X] = [BF_4]$^-$, [PF_6]$^-$), and alcohols for the synthesis of dialkyl carbonates under mild conditions. During the process, the IL served as both a green solvent with low toxicity and also generated NHC. This eliminated the use of toxic organic solvents, and the addition of supporting electrolytes while favoring good conversion, selectivity. Since then there were numerous works published in this area which focused on the transformation of CO_2 via the electrosynthesis route to the cyclic carbonates such as propylene carbonate using Lewis acid ILs, ILs with hydrogen bond donors. The high viscosities of ILs, combined with their low CO_2 conversion, limit their direct application. This has motivated researchers to extend the studies to binary IL systems using either water or an organic solvent as a co-solvent. Functionalizing the ILs or changing the IL network with –OH or –COOH via hydrogen bonding was demonstrated to have a greater influence on the electrosynthesis process than unmodified ILs. Carbamates, carboxylates and anilines have significant attention as a value-added chemical for applications ranging from pharmaceuticals, agro chemicals, dyes, perfumes, intermediates for detergents and so on. Producing them via the electrosynthesis can be a beneficial way in terms of environment friendliness, moderate reaction conditions, eliminating the use of a toxic compound phosgene. To produce these compounds, it is necessary to transform the CO_2 and its corresponding substrate, create either a C–N bond using amines (for carbamates) or C–C (for carboxylates). Table 4 lists an overview of the various organic compounds produced by electrosynthesis in CO_2-saturated ILs along with the substrates, reactor types, catalysts employed in the process.

Carbonates (cyclic/dialkyl) are produced when CO_2 is coupled to an alcohol during electrosynthesis, while coupling to an amine (aliphatic/aromatic) yields carbamates. Electrocarboxylation involves the coupling of CO_2 to a radical/anion produced during the electrochemical reduction of organic halides (alkyl/aryl), thereby yielding carboxylate derivative. Feng's group [66] performed the electrolysis at 50 °C in an undivided cell with Pt cathode and Mg anode at certain concentrations of acetophenone-an aromatic ketone, in [BMIm][BF_4] saturated with CO_2. This was carried out in the presence of the alkylating agent, methyl iodide (CH_3I) to afford the corresponding α-hydroxycarboxylic acid methyl ester with yields of ~56–62%. The corresponding alcohols were obtained as the main by-products. Zhao et al. [20] investigated the effect of proton availability in ILs on the product distribution of acetophenone during electrocarboxylation with CO_2. They observed that dry pyrolidinum-based IL [BMPyl][Tf_2N] with limited proton availability was an appropriate medium for their electrocarboxylation system, yielding 98% 2-hydroxy-2-phenylpropionic acid. The competing reactions are not beneficial to the electrocarboxylation and some studies suggested that the product distribution is strongly dependant on the medium. Lu et al [58] studied the formation of phenylacetic acid in CO_2-saturated [BMIm][BF_4] via electrocarboxylation of benzyl chloride with silver cathode as the catalyst. Initially, benzyl chloride was electroreduced to its corresponding radical and subsequently coupled to CO_2 to yield the carboxylate derivatives. Hiejima et al. [59] reported the synthesis of α-chloroethylbenzene via electrocarboxylation in [DEME][Tf_2N]-based ionic liquid compressed with CO_2. The experiments were carried out using a Pt cathode and a Mg anode at various temperatures and pressures; but obtained poor carboxylic acid product yields (~20%). Atobe et al [67] improved the yield further for the same process to 50% by using supercritical CO_2 in [DEME][Tf_2N]-based IL. However, obtaining the products via the transformation approach is a greener and easier option because it:

- avoids the use of hazardous reagents such as phosgene and cyanide
- simplifies the purification of the resultant products utilising single step chromatographic separations.

Table 4. Summary of IL-based electrolytes used in the electrochemical conversion of CO_2 via the transformation route.

Product Family	Product	Electrolyte	Substrates	Cathode	Anode	Reactor Type	Reference
Dialkyl carbonate	Dimethyl carbonate	[BMIm][Br]	Methanol and propylene oxide	Platinum	Platinum	One compartment cell	[16]
Carboxylate	2–hydroxy–2–phenyl propionic acid	[BMM'Im][BF$_4$]	Acetophenone	Glassy carbon (cylindrical tube)	Magnesium	Undivided cell	[20]
	2–hydroxy–2–phenyl propionic acid	[BMPy][Tf$_2$N]	Acetophenone	Glassy carbon (cylindrical tube)	Magnesium	Undivided cell	
Cyclic carbonate	Styrene carbonate	[BMIm][BF$_4$]	Styrene, glycol, methyl iodide (alkylating agent), potassium carbonate	Titanium	Platinum spiral	Two compartment cell divided by a cation exchange membrane	[55]
Dialkyl carbonate	Dimethyl carbonate	[B'MIm][Cl]	Methanol	Graphite	Platinum	Undivided cell (four neck bottle)	[57]
Carboxylate	Phenyl acetic acid	[BMIm][BF$_4$]	Benzyl chloride	Silver cylinder	Magnesium	Undivided cell	[58]
Carboxylate	2–phenyl propionic acid	[DEME][Tf$_2$N]	α–chloroethyl benzene	Platinum plate	Magnesium	High pressure vessel	[59]
Dialkyl carbonate	Dimethyl carbonate	[BMIm][BF$_4$]	Methanol; methyl iodide (alkylating agent)	Silver-coated nanoporous copper	Platinum foil	Undivided cell	[60]
Carbamate	6–amino nicotinic acid	[BMIm][BF$_4$]	2–amino–5–bromopyridine	Silver	Magnesium rod	Undivided cell	[61]
Carbamate	Organic carbamates	[BMIm][BF$_4$]	Amines, O$_2$, ethyl iodide (alkylating agent)	Copper	Platinum spiral	Two compartment 3-electrode cell	[62]
Dialkyl carbonate	Dimethyl carbonate	[AMIm][Br]	Methanol	Graphite	Platinum	Undivided cell (four neck bottle)	[63]
Dialkyl carbonate	Dimethyl carbonate	[BMIm][Br]	Potassium ethoxide	Platinum/ niobium plates	-	Divided electrochemical cell	[64]
Dialkyl carbonate	Dimethyl carbonate	[BMIm][Br]	Potassium ethoxide	Graphite	-	Divided electrochemical cell	[65]

3. Techno-Commercial Challenges and Future Road Map

A large array of ILs in the range of ~10^{18} were discovered through different arrangements of cations and anions, which can possibly be synthetized and present a good platform for design [68–70], but, it is interesting to note that only ~10^2–10^3 ILs were commercialised. This could be ascribed to the complexities in the methods of preparation, expensive equipment, controlled operating conditions, availability of raw materials, costs, storage and safe handling after the synthesis, and limited availability of the physical, chemical property data base [71–76]. Most of the properties available so far are derived using the machine learning methods [77]. In addition, studies and data concerning the environmental considerations, recycling, biodegradability, recovery and reuse, thermal stability is very limited and lacks significant consideration [78]. Also, the high viscosities of traditional ILs limit their direct application for the electrosynthesis of organic molecules from CO_2, electroreduction of CO_2 to CO and other chemicals, which is the main reason why binary IL systems have received a lot of significant interest recently. The higher viscosity of ILs pose the difficulties in transportation of the reactant species (CO_2) despite their high solubility. As a result, from the standpoint of functioning in a real-world application, a simple and efficient system is still required. In the recent times, DESs prepared by mixing imidazolium chloride ILs with hydrogen bond donors such as [ChCl:Urea], [ChCl:EG], 1M [ChCl] in EG, and

[BMIm][Cl]:EG and, were shown to exhibit substantially lower viscosities than expected when compared to unmodified ILs [17,45]. Figure 5 compares the viscosities of the ILs and DES systems employed for either electroreduction or electro-organic transformation utilising CO_2.

The selected anion will have a known influence on the viscosity of the ionic liquid as well as the transport of the species to the electrode surface. To demonstrate the influence of anion, selected data on viscosity of the systems that have been commercialized, developed, are gathered and plotted as shown in Figure 5. In a specific case where the anion is paired with the same cation: $[BMIm]^+$, the effect of the anion on viscosity follows the trend: $[DCA]^- < [Br]^- < [Tf_2N]^- < [TfO]^- < [BF_4]^- < [PF_6]^- < [Ac]^-$. Certain trends from this data may be extrapolated, that hold true regardless of cation's identity. For instance, results from Fomin et al. [21], Crosthwaite et al. [79] demonstrated that increasing the cationic chain length with common anion $[Tf_2N]^-$ increases the viscosity up to 9000 mPa s. Amongst the literature reported so far, indicated that imidazolium-based ILs are the most studied ILs for electroreduction of CO_2 electrosynthesis of organic molecules using CO_2 as reaction media due to their high CO_2 capture ability, Lewis acid-base interaction [10,14,19,22,36], but the hygroscopic nature of imidazolium cations combined with their high viscosities not only decreases the mass transportation of the reactant species but affect the reduction, electrosynthesis. The majority of studies involving the direct electrochemical conversion of CO_2 to dialkyl carbonates (such as DMCs) are conducted in imidazolium-based ionic liquids (ILs): $[BMIm][BF_4]$ and $[EMIm][BF_4]$ [44]. These solvents are demonstrated to be promising because of the high CO_2 solubility compared to conventional solvents [0.14, 0.10 vs 0.09 (acetonitrile), 0.07 (ethanol)] at 300 K and 10 bar [51].

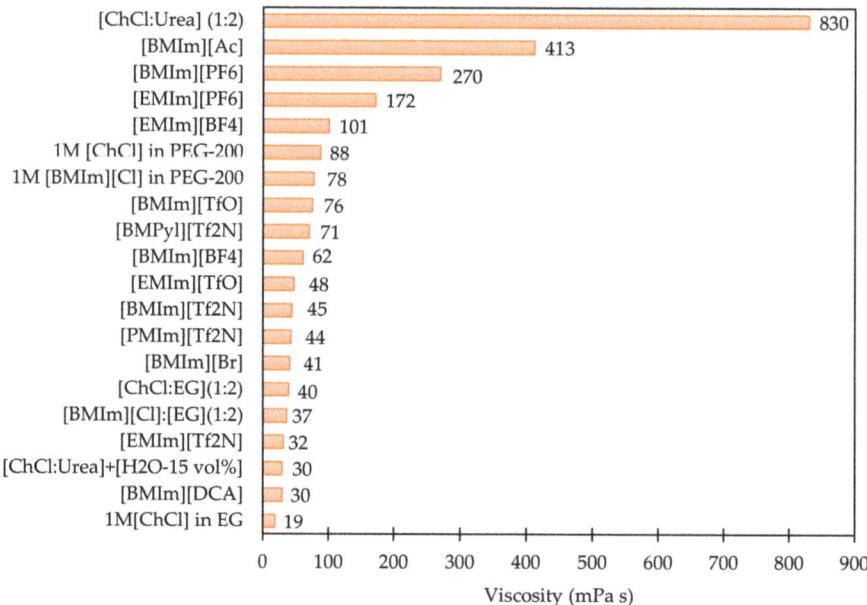

Figure 5. Plot comparing the viscosity of the ILs, DESs that are used in electrochemical conversion of CO_2 at 298 K.

Despite these promising yields for both the processes, these systems utilise carcinogenic substances such as $[BF_4]^-$, $[PF_6]^-$ for electroreduction whereas alkylating agents such as methyl iodide and other harmful compounds such as propylene oxide are used for the electrosynthesis of organic carbonates. Subsequently significant efforts were made to eliminate the use of such harmful compounds and reduce the number of production steps. Compton et al. [22] explored the electroreduction of CO_2 in [BMIm][Ac] which exhibited a high CO_2 solubility of 1520 mM. The CO_2 in [BMIm][Ac] underwent a chemically irreversible, one-electron transfer to the radical anion ($CO_2\bullet^-$), and probably favor the formation of oxalate, CO, and carbonate. Yuan et al. [16,57,80] made a series of attempts to substitute the harmful methyl iodide with a variety of basic compounds: CH_3ONa, NaOH, CH_3OK, K_2CO_3, KOH, to electrosynthesise DMC via CO_2 transformation on Pt electrodes in dialkyl imidazolium ILs. The results demonstrated that dialkyl imidazolium ILs – (CH_3OK)–(CH_3OH) system displayed the highest yield. The authors observed that the higher K^+ ion interaction with adsorbed CO_2 resulted in better stabilisation of the CO_2^- anion which favoured the electrosynthesis pathway for the formation of dimethyl carbonates. While carcinogenic chemicals can be avoided, the use of less hazardous ILs has certain limitation such as high cost and viscosity besides low cost. For instance, the viscosity of [BMIm][Ac] is very high (413 mPa) in the case of CO_2 electroreduction, while the DMC yield achieved with CH_3OK as alkylating agent was ~4%. Hence, a more cost-efficient and feasible system for the electrochemical reduction of CO_2 and electrochemical conversions involving CO_2 to respective carbonates, carbamates, carboxylates and organic compounds needs to be developed.

Evaluating the electrochemical conversion of CO_2 at higher temperatures in the range of 100 °C employing newly designed ILs, DES systems can result in industrially relevant rates of CO_2 conversions to ethanol and ethylene besides CO, HCOOH. Such a temperature is expected to decrease the viscosity of the ILs while enhancing the mobility of the reactant species with increased solubility. As a result, high Faradaic efficiencies and current densities can be reproducibly achieved. With key advances in catalysts (2D structures, heteroatom doped structures, single faceted crystal structures) that lead to impressive performance at the lab scale, additional work on these catalysts in newly developed less hazardous non-aqueous systems is required to provide benchmarks against which industries can compare their results. Because ILs are widely acknowledged as one of the most expensive compounds, their cycle stability in the electrochemical reduction process should be taken into account. In addition, the best experimental findings should be obtained at the lowest feasible cost. [EMIm][BF_4] performed better in electroreduction of CO_2 to CO and other value-added compounds in several imidazolium-based ILs. However, potential issues arise when [EMIm][BF_4] is hydrolyzed, which releases HF and certain anions): [BF_3OH]$^-$, [$BF_2(OH)_2$]$^-$, [$BF(OH)_3$]$^-$,and [$B(OH)_4$]$^-$. This was shown to occur upon addition of water (which is introduced to compensate for viscosity) [36]. The formation of HF makes the CO_2 saturated solution more acidic and aggressive, corroding the equipment and electrodes, whilst the other complexes increase the reaction rate. Also, hydrolysis of [BMIm][BF_4] makes the recycling difficult owing to its reduced stability and as a consequence, increases the experimental cost.

To solve this problem, [BF_4]$^-$ anions are replaced with [TfO]$^-$ and [BMIm]$^+$ as the cation as an alternative [81]. While the imidazolium cations in ILs play a key role in the electrochemical conversion of CO_2 (reduction, transformation), the anions have an impact on the pricing of imidazolium-based ILs. Currently, there are numerous imidazolium based ILs that exist commercially with different properties and variable costs. Also, there were new developments in DESs as an electrolyte or reaction media for the electrochemical conversion of CO_2. The high cost associated with ILs is one of the bottlenecks that hinder their industrial use. Hence, it is critical to evaluate and understand the price and economic feasibility of ILs. Besides, with the growing significant interest in DESs it is worth to compare their prices to understand the economic impacts better.

Figure 6 depicts the plots based on commercially accessible pricing, and Figure 7 is plotted based on the cost of the raw materials utilized in the synthesis of these ILs, DESs and the methodology adopted by Cui et al. [15] in their review. The cost for DESs is evaluated based on their mole ratios of the individual components and the commercially available pricing. Since the price variations per kg of the product are quite significant, the associated costs with ILs, DESs synthesized from raw materials are presented individually.

Data from Figures 6 and 7 signify that the cost of ILs paired with [TfO]⁻ or [Tf₂N]⁻ anion is higher than that of other ILs. These cost comparison data derived from the Figures 6 and 7 validates the fact that the cost of imidazolium-based ILs is mostly determined by the anions, in line with the above analysis. It is worth noting that the price of DESs from Figures 6 and 7 are significantly lower than ILs, which further validates the scientific claims. Also, it might be the primary reason to motivate the industries to focus on their development.

By comparing and evaluating the market pricing of ILs, DESs, it is clear that there exists a significant gap between the price of commercially marketed ILs, DESs plotted in Figure 6 and the price of ILs, DESs produced from raw materials represented in Figure 7. This suggests that increasing the production of ILs, DESs through large scale preparation for the electrochemical conversion of CO_2 will offer a competitive cost benefit compared with the commercial market pricing, making the possibility to achieve the price of around $1 per kilogram of IL (or) DES [71].

While there are significant efforts in reducing the costs for the conversion processes either by employing co-solvents such as acetonitrile, alcohols or using mixed electrolytes (ILs + cosolvent), the price is still competitive. Therefore, prior attention needs to be paid for their applications in large scale industries. Also, the current commercially available ILs that are employed for the electrochemical conversion of CO_2 contain carcinogenic, harmful or toxic substances. These substances might need to be replaced with less hazardous ones such as acetate in order for them to comply with the EU regulations and such as Registration, Evaluation, Authorisation and Restriction of Chemicals (REACH).

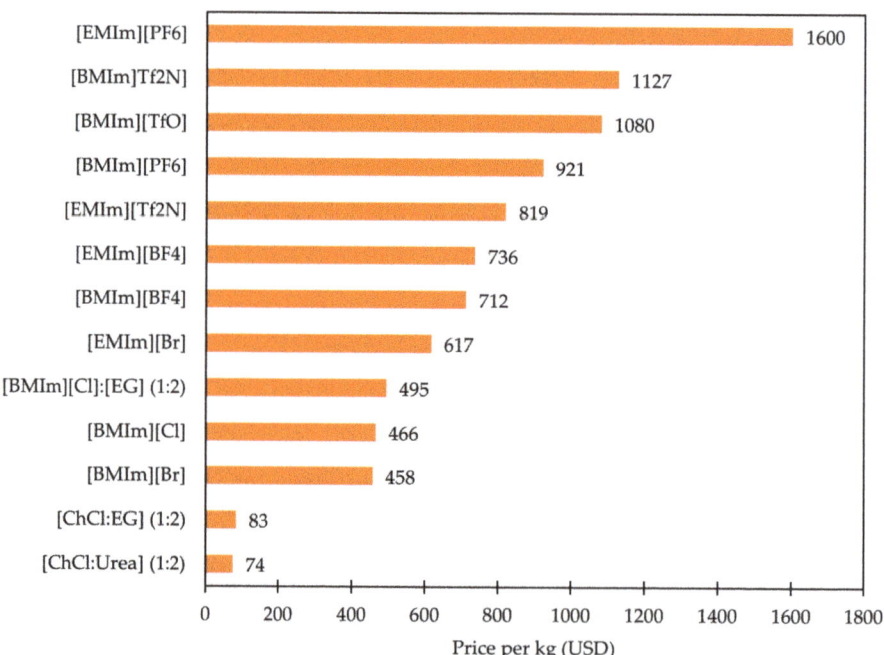

Figure 6. Plot comparing the cost of ILs, DESs in USD per kg based on commercial pricing.

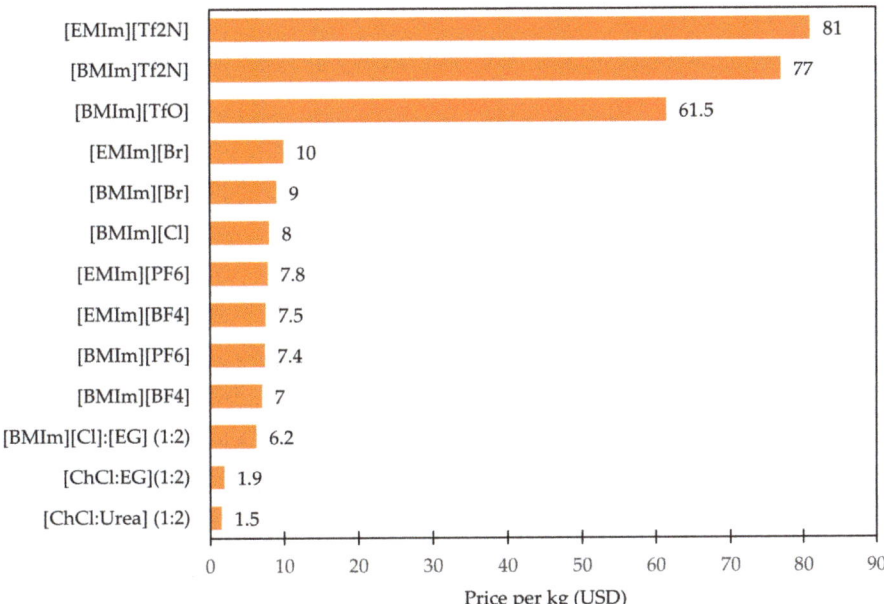

Figure 7. Plot comparing the cost of the ILs, DESs in USD per kg that are computed based on the raw materials cost by adopting the methodology reported by [15].

Besides the predominantly expensive ILs, another commercial challenge is the costs associated with electricity which influences the profitability of the electrochemical processes that utilise CO_2 either in the electroreduction or electrosynthesis (of organic molecules). Techno-economic model results based on the generalised electrochemical reduction plant developed by Jouny et al. [82] demonstrates that electricity costs influence the net present value significantly. Besides, it also demanded for a continuous supply of cheap electricity to reduce the overall commercial electrochemical CO_2 reduction plant production costs. One strategy for lowering the production cost would be the use of electricity generation via renewable energy source such as solar. The utilisation of renewable energy source to produce electricity can be considered as an attractive approach to produce carbon feedstocks such as hydrocarbons (formic acid, ethylene) in a carbon-neutral way via the electrochemical routes (reduction, synthesis). Utilising the renewable energy resources such as wind, solar, geothermal energies for the CO_2 conversion via the electrochemical route can reduce the global weighted-average levelized cost of electricity (LCOE). As per the IRENA reports, solar energy was shown to be the low-cost option that can be used to provide both electrical and thermal energy [83]. Using the solar energy as a source of renewable electricity for the conversion of CO_2 to produce renewable sources is expected to further reduce the LCOE, production costs and contribute to the carbon neutral economy. Finally, the materials of construction of the suitable equipment used for the electrochemical conversion of CO_2 needs to be chosen wisely when choosing certain corrosive ILs such as imidazolium-based chlorides. Identifying the alternatives to the commercial hazardous anions and the replacement of imidazolium cations with other family of ILs will be a beneficial approach for rapid commercialization of non-aqueous systems for electrochemical conversion of CO_2. Similarly, performing the real-time study with the already developed DESs for CO_2 conversion technologies will provide a detailed insight into the complexities that might be encountered during their commercialization.

While the development of electrochemical conversion of CO_2 processes seeks significant capital investments, performing the scale-up of electrochemical reactions at a scale spanning between 100–1000 L with the ILs, IL-based mixed electrolytes makes it expensive. Also, the hygroscopic, corrosive nature of the imidazolium chloride-based ILs demands sophisticated equipment to operate and produce them in bulk. There is an urgent need to increase the CO_2 utilisation to produce products for renewable energy technologies via the electrochemical conversions such as reduction, synthesis (through transformation) while optimising the costs to meet the carbon neutral economy. Furthermore, their compatibility with co-solvents such as water with better recycling characteristics and optimisation of the process intensification technologies will not only reduce the operational cost but also the equipment, overheads.

4. Future Perspective

Most of the works covered in this review focused on the combinations of imidazolium, pyrrolidinium cations with different anions: $[BF_4]^-$, $[PF_6]^-$, $[TfO]^-$, $[Tf_2N]^-$, $[FAP]^-$ with concentrations from pure ILs to millimolar, molar ranges but the optimum compositions were not determined. As these systems pose certain safety hazards such as carcinogenicity, toxicity, bio-degradability, they might have to be replaced with less hazardous category of ILs in order to comply with the safety regulations such as REACH, Restriction of Hazardous Substances (RoHS). Development of halide-free ILs, DESs alternative to imidazolium chlorides (such as acetates), less hazardous alkylating agents (eg.CH_3OK) can be considered as a way forward.

Few works on the application of halide free IL systems such as acetate-based, and DESs such as [ChCl:Urea/EG], imidazolium chloride-based ones have been investigated as potential alternative electrolytes for the electroreduction of CO_2 [17], but studies on the electrosynthesis of organic compounds through the transformation route are still scarce. Also, most of the works focused on the electrosynthesis of carbonates with little on carboxylates, carbamates employing CO_2-saturated ILs.

Hence, a more cost-effective system to convert CO_2 to the respective organic compounds such as CO, HCOOH, ethylene (C_2H_4), and carbonates, carboxylates, carbamates through electrochemical approach needs to be developed. Identification of the reaction intermediates through mechanistic understanding during the electrochemical conversion of CO_2 in either ways will be the key challenge [9,15,50]. However: (i) enhancing the real time performance of low-cost halide free ILs (such as acetate), and imidazolium chloride-based DESs, (ii) accelerating the research on ILs' recovery and reuse will be a significant future challenge.

Preliminary screening of the alternatives based on their properties (physical, chemical, electrochemical), safety, cost from a variety of ILs, DESs could reduce the time and efforts in identifying the best ones. Utilising such electrolytes for either of the electrochemical conversions: CO_2 reduction, transformation via synthesis will help to increase the performance of CO_2 utilization technologies besides promoting the scope of ILs/DESs in the renewable energy sector.

5. Conclusions

The present review summarised the developments in electrochemical conversion of CO_2 value added chemicals through the electroreduction and electroorganic transformation using ILs, DESs as the electrolyte. Imidazolium and pyrrolidinium cations have been proved to be very effective for enhancing the performance of the catalysts towards the electroreduction of CO_2, with imidazolium-based ILs being predominant. While the cations of ILs play a multifunctional role in the electroreduction system, interacting with reaction intermediates and possibly acting as a co-catalyst, the anions contribute to electrolyte stability but increase the overall cost. The studies referenced reflect the scientific community's significant attempts to produce efficient electrochemical CO_2 conversions, either through reduction or transformation.

Having understood the progress in the electrochemical conversion of CO_2 employing ILs, DESs, the techno-commercial challenges, the non-aqueous electrolyte based approach could be considered as a promising way forward for obtaining the resultant products (HCOOH, HCHO, C_2H_4 etc.). These products serve as a renewable source to produce value added chemicals at industrial scale. Clearly, significant improvements are desired in terms of identifying the less hazardous ILs, scope of DESs for electrochemical conversion of CO_2 to make a step towards establishing such technology on a larger scale and demonstrating it as a sustainable process. While substantial improvements have been recorded in understanding the mechanistic aspects of CO_2 electrochemical transformations in ILs, investigations using DESs as the electrolyte deserve considerable research.

The primary challenges to focus at this stage have been outlined in this review which is expected to potentially support the research community working in this field and aid with rapid commercialisation of ILs/DESs.

Author Contributions: Conceptualization, K.K.M. and S.P.; writing —original draft preparation, K.K.M.; writing—review and editing, S.P.; Supervision, S.P. All authors have read and agreed to the published version of the manuscript.

Funding: This research has received funding from the European Union's Horizon 2020 research and innovation program under the Marie Sklodowska-Curie grant agreement No. 885793. A full fee waiver APC was granted for this article to be published in *Materials*.

Institutional Review Board Statement: Not applicable.

Informed Consent Statement: Not applicable.

Data Availability Statement: The data presented in this study are available upon request from the corresponding author.

Conflicts of Interest: The authors declare no conflict of interest.

References

1. Song, C. Global challenges and strategies for control, conversion and utilization of CO_2 for sustainable development involving energy, catalysis, adsorption and chemical processing. *Catal. Today* **2006**, *115*, 2–32. [CrossRef]
2. Hospital-Benito, D.; Lemus, J.; Moya, C.; Santiago, R.; Ferro, V.R.; Palomar, J. Techno-economic feasibility of ionic liquids-based CO_2 chemical capture processes. *Chem. Eng. J.* **2021**, *407*, 127196. [CrossRef]
3. Li, F.; MacFarlane, D.R.; Zhang, J. Recent advances in the nanoengineering of electrocatalysts for CO_2 reduction. *Nanoscale* **2018**, *10*, 6235–6260. [CrossRef] [PubMed]
4. Global Energy Review. Available online: https://www.iea.org/reports/global-energy-review-2020/renewables (accessed on 26 May 2021).
5. Chen, C.; Khosrowabadi Kotyk, J.F.; Sheehan, S.W. Progress toward Commercial Application of Electrochemical Carbon Dioxide Reduction. *Chem* **2018**, *4*, 2571–2586. [CrossRef]
6. Gomes, J.F.P. *Carbon Dioxide Capture and Sequestration: An Integrated Overview of Available Technologies*; Gomes, J.F.P., Ed.; Nova Science Pub Inc.: New York, NY, USA, 2013.
7. Zhang, S.; Sun, J.; Zhang, X.; Xin, J.; Miao, Q.; Wang, J. Ionic liquid-based green processes for energy production. *Chem. Soc. Rev.* **2014**, *43*, 7838–7869. [CrossRef] [PubMed]
8. Pollok, D.; Waldvogel, S.R. Electro-organic synthesis — a 21 st century technique. *Chem. Sci.* **2020**, *11*, 12386–12400. [CrossRef] [PubMed]
9. Anastasiadou, D.; Hensen, E.J.M.; Figueiredo, M.C. Electrocatalytic synthesis of organic carbonates. *Chem. Commun.* **2020**, *56*, 13082–13092. [CrossRef]
10. Vasilyev, D.V.; Dyson, P.J. The Role of Organic Promoters in the Electroreduction of Carbon Dioxide. *ACS Catal.* **2021**, *11*, 1392–1405. [CrossRef]
11. Mehrkesh, A.; Karunanithi, A.T. Optimal design of ionic liquids for thermal energy storage. *Comput. Chem. Eng.* **2016**, *93*, 402–412. [CrossRef]
12. Li, X.; Wang, S.; Li, L.; Sun, Y.; Xie, Y. Progress and Perspective for in Situ Studies of CO_2 Reduction. *J. Am. Chem. Soc.* **2020**, *142*, 9567–9581. [CrossRef] [PubMed]
13. Alvarez-Guerra, M.; Albo, J.; Alvarez-Guerra, E.; Irabien, A. Ionic liquids in the electrochemical valorisation of CO_2. *Energy Environ. Sci.* **2015**, *8*, 2574–2599. [CrossRef]
14. Garg, S.; Li, M.; Weber, A.Z.; Ge, L.; Li, L.; Rudolph, V.; Wang, G.; Rufford, T.E. Advances and challenges in electrochemical CO_2 reduction processes: An engineering and design perspective looking beyond new catalyst materials. *J. Mater. Chem. A* **2020**, *8*, 1511–1544. [CrossRef]

15. Cui, Y.; He, B.; Liu, X.; Sun, J. Ionic Liquids-Promoted Electrocatalytic Reduction of Carbon Dioxide. *Ind. Eng. Chem. Res.* **2020**, *59*, 20235–20252. [CrossRef]
16. Yuan, D.D.; Yuan, B.B.; Song, H.; Niu, R.X.; Liu, Y.X. Electrochemical Fixation of Carbon Dioxide for Synthesis Dimethyl Carbonate in Ionic Liquid BMimBr. *Adv. Mater. Res.* **2014**, *953–954*, 1180–1183. [CrossRef]
17. Vasilyev, D.V.; Rudnev, A.V.; Broekmann, P.; Dyson, P.J. A General and Facile Approach for the Electrochemical Reduction of Carbon Dioxide Inspired by Deep Eutectic Solvents. *ChemSusChem* **2019**, *12*, 1635–1639. [CrossRef]
18. Tanner, E.E.L.; Batchelor-McAuley, C.; Compton, R.G. Carbon Dioxide Reduction in Room-Temperature Ionic Liquids: The Effect of the Choice of Electrode Material, Cation, and Anion. *J. Phys. Chem. C* **2016**, *120*, 26442–26447. [CrossRef]
19. Sun, X.; Kang, X.; Zhu, Q.; Ma, J.; Yang, G.; Liu, Z.; Han, B. Very highly efficient reduction of CO_2 to CH_4 using metal-free N-doped carbon electrodes. *Chem. Sci.* **2016**, *7*, 2883–2887. [CrossRef]
20. Zhao, S.-F.; Horne, M.; Bond, A.M.; Zhang, J. Electrocarboxylation of acetophenone in ionic liquids: The influence of proton availability on product distribution. *Green Chem.* **2014**, *16*, 2242–2251. [CrossRef]
21. Fomin, A. Studies on the Electrodeposition of Aluminium from Different Air and Water Stable Ionic Liquids. Ph.D. Thesis, Clausthal University of Technology, Clausthal-Zellerfeld, Germany, 2010. Available online: https://d-nb.info/1008342475/34 (accessed on 28 April 2021).
22. Barrosse-Antle, L.E.; Compton, R.G. Reduction of carbon dioxide in 1-butyl-3-methylimidazolium acetate. *Chem. Commun.* **2009**, 3744. [CrossRef]
23. Srivastava, R.; Nouseen, S.; Chattopadhyay, J.; Woi, P.M.; Son, D.N.; Bastakoti, B.P. Recent Advances in Electrochemical Water Splitting and Reduction of CO2 into Green Fuels on 2D Phosphorene-Based Catalyst. *Energy Technol.* **2021**, *9*, 1–19. [CrossRef]
24. Yang, C.H.; Nosheen, F.; Zhang, Z.C. Recent progress in structural modulation of metal nanomaterials for electrocatalytic CO_2 reduction. *Rare Met.* **2021**, *40*, 1412–1430. [CrossRef]
25. König, M.; Vaes, J.; Klemm, E.; Pant, D. Solvents and Supporting Electrolytes in the Electrocatalytic Reduction of CO_2. *iScience* **2019**, *19*, 135–160. [CrossRef] [PubMed]
26. Weng, L.-C.; Bell, A.T.; Weber, A.Z. Towards membrane-electrode assembly systems for CO_2 reduction: A modeling study. *Energy Environ. Sci.* **2019**, *12*, 1950–1968. [CrossRef]
27. Kaneco, S.; Iiba, K.; Katsumata, H.; Suzuki, T.; Ohta, K. Electrochemical reduction of high pressure CO_2 at a Cu electrode in cold methanol. *Electrochim. Acta* **2006**, *51*, 4880–4885. [CrossRef]
28. Rosen, B.A.; Salehi-Khojin, A.; Thorson, M.R.; Zhu, W.; Whipple, D.T.; Kenis, P.J.A.; Masel, R.I. Ionic Liquid-Mediated Selective Conversion of CO_2 to CO at Low Overpotentials. *Science* **2011**, *334*, 643–644. [CrossRef] [PubMed]
29. Zhao, G.; Jiang, T.; Han, B.; Li, Z.; Zhang, J.; Liu, Z.; He, J.; Wu, W. Electrochemical reduction of supercritical carbon dioxide in ionic liquid 1-n-butyl-3-methylimidazolium hexafluorophosphate. *J. Supercrit. Fluids* **2004**, *32*, 287–291. [CrossRef]
30. Rosen, B.A.; Haan, J.L.; Mukherjee, P.; Braunschweig, B.; Zhu, W.; Salehi-Khojin, A.; Dlott, D.D.; Masel, R.I. In Situ Spectroscopic Examination of a Low Overpotential Pathway for Carbon Dioxide Conversion to Carbon Monoxide. *J. Phys. Chem. C* **2012**, *116*, 15307–15312. [CrossRef]
31. Asadi, M.; Kumar, B.; Behranginia, A.; Rosen, B.A.; Baskin, A.; Repnin, N.; Pisasale, D.; Phillips, P.; Zhu, W.; Haasch, R.; et al. Robust carbon dioxide reduction on molybdenum disulphide edges. *Nat. Commun.* **2014**, *5*, 4470. [CrossRef] [PubMed]
32. Asadi, M.; Kim, K.; Liu, C.; Addepalli, A.V.; Abbasi, P.; Yasaei, P.; Phillips, P.; Behranginia, A.; Cerrato, J.M.; Haasch, R.; et al. Nanostructured transition metal dichalcogenide electrocatalysts for CO_2 reduction in ionic liquid. *Science* **2016**, *353*, 467–470. [CrossRef]
33. Kumar, B.; Asadi, M.; Pisasale, D.; Sinha-Ray, S.; Rosen, B.A.; Haasch, R.; Abiade, J.; Yarin, A.L.; Salehi-Khojin, A. Renewable and metal-free carbon nanofibre catalysts for carbon dioxide reduction. *Nat. Commun.* **2013**, *4*, 2819. [CrossRef]
34. Boddien, A.; Loges, B.; Gärtner, F.; Torborg, C.; Fumino, K.; Junge, H.; Ludwig, R.; Beller, M. Iron-Catalyzed Hydrogen Production from Formic Acid. *J. Am. Chem. Soc.* **2010**, *132*, 8924–8934. [CrossRef]
35. Chu, D.; Qin, G.; Yuan, X.; Xu, M.; Zheng, P.; Lu, J. Fixation of CO_2 by Electrocatalytic Reduction and Electropolymerization in Ionic Liquid–H_2O Solution. *ChemSusChem* **2008**, *1*, 205–209. [CrossRef]
36. Rosen, B.A.; Zhu, W.; Kaul, G.; Salehi-Khojin, A.; Masel, R.I. Water Enhancement of CO_2 Conversion on Silver in 1-Ethyl-3-Methylimidazolium Tetrafluoroborate. *J. Electrochem. Soc.* **2013**, *160*, H138–H141. [CrossRef]
37. Salehi-Khojin, A.; Jhong, H.-R.M.; Rosen, B.A.; Zhu, W.; Ma, S.; Kenis, P.J.A.; Masel, R.I. Nanoparticle Silver Catalysts That Show Enhanced Activity for Carbon Dioxide Electrolysis. *J. Phys. Chem. C* **2013**, *117*, 1627–1632. [CrossRef]
38. Feroci, M.; Chiarotto, I.; Ciprioti, S.V.; Inesi, A. On the reactivity and stability of electrogenerated N-heterocyclic carbene in parent 1-butyl-3-methyl-1H-imidazolium tetrafluoroborate: Formation and use of N-heterocyclic carbene-CO_2 adduct as latent catalyst. *Electrochim. Acta* **2013**, *109*, 95–101. [CrossRef]
39. Zhou, F.; Liu, S.; Yang, B.; Wang, P.; Alshammari, A.S.; Deng, Y. Highly selective electrocatalytic reduction of carbon dioxide to carbon monoxide on silver electrode with aqueous ionic liquids. *Electrochem. Commun.* **2014**, *46*, 103–106. [CrossRef]
40. Quezada, D.; Honores, J.; García, M.; Armijo, F.; Isaacs, M. Electrocatalytic reduction of carbon dioxide on a cobalt tetrakis(4-aminophenyl)porphyrin modified electrode in BMImBF 4. *New J. Chem.* **2014**, *38*, 3606–3612. [CrossRef]
41. Martindale, B.C.M.; Compton, R.G. Formic acid electro-synthesis from carbon dioxide in a room temperature ionic liquid. *Chem. Commun.* **2012**, *48*, 6487. [CrossRef] [PubMed]

42. Vedharathinam, V.; Qi, Z.; Horwood, C.; Bourcier, B.; Stadermann, M.; Biener, J.; Biener, M. Using a 3D Porous Flow-Through Electrode Geometry for High-Rate Electrochemical Reduction of CO_2 to CO in Ionic Liquid. *ACS Catal.* **2019**, *9*, 10605–10611. [CrossRef]
43. Bruzon, D.A.; Tiongson, J.K.; Tapang, G.; Martinez, I.S. Electroreduction and solubility of CO_2 in methoxy- and nitrile-functionalized imidazolium (FAP) ionic liquids. *J. Appl. Electrochem.* **2017**, *47*, 1251–1260. [CrossRef]
44. García, G.; Aparicio, S.; Ullah, R.; Atilhan, M. Deep Eutectic Solvents: Physicochemical Properties and Gas Separation Applications. *Energy Fuels* **2015**, *29*, 2616–2644. [CrossRef]
45. Verma, S.; Lu, X.; Ma, S.; Masel, R.I.; Kenis, P.J.A. The effect of electrolyte composition on the electroreduction of CO_2 to CO on Ag based gas diffusion electrodes. *Phys. Chem. Chem. Phys.* **2016**, *18*, 7075–7084. [CrossRef]
46. Bohlen, B.; Wastl, D.; Radomski, J.; Sieber, V.; Vieira, L. Electrochemical CO_2 reduction to formate on indium catalysts prepared by electrodeposition in deep eutectic solvents. *Electrochem. Commun.* **2020**, *110*, 106597. [CrossRef]
47. Garg, S.; Li, M.; Rufford, T.E.; Ge, L.; Rudolph, V.; Knibbe, R.; Konarova, M.; Wang, G.G.X. Catalyst–Electrolyte Interactions in Aqueous Reline Solutions for Highly Selective Electrochemical CO_2 Reduction. *ChemSusChem* **2019**, 304–311. [CrossRef]
48. Ahmad, N.; Wang, X.; Sun, P.; Chen, Y.; Rehman, F.; Xu, J.; Xu, X. Electrochemical CO_2 reduction to CO facilitated by MDEA-based deep eutectic solvent in aqueous solution. *Renew. Energy* **2021**, *177*, 23–33. [CrossRef]
49. Sakakura, T.; Choi, J.-C.; Yasuda, H. Transformation of Carbon Dioxide. *Chem. Rev.* **2007**, *107*, 2365–2387. [CrossRef] [PubMed]
50. Tan, Y.; Sun, X.; Han, B. Ionic Liquid-Based electrolytes for CO_2 electroreduction and CO_2 electroorganic transformation. *Natl. Sci. Rev.* **2021**. [CrossRef]
51. Kathiresan, M.; Velayutham, D. Ionic liquids as an electrolyte for the electro synthesis of organic compounds. *Chem. Commun.* **2015**, *51*, 17499–17516. [CrossRef]
52. Cai, X.; Xie, B. Direct Carboxylative Reactions for the Transformation of Carbon Dioxide into Carboxylic Acids and Derivatives. *Synthesis* **2013**, *45*, 3305–3324. [CrossRef]
53. North, M.; Pasquale, R.; Young, C. Synthesis of cyclic carbonates from epoxides and CO_2. *Green Chem.* **2010**, *12*, 1514. [CrossRef]
54. Yang, H.; Gu, Y.; Deng, Y.; Shi, F. Electrochemical activation of carbon dioxide in ionic liquid: Synthesis of cyclic carbonates at mild reaction conditions. *Chem. Commun.* **2002**, 274–275. [CrossRef]
55. Wu, L.-X.; Wang, H.; Xiao, Y.; Tu, Z.-Y.; Ding, B.-B.; Lu, J.-X. Synthesis of dialkyl carbonates from CO_2 and alcohols via electrogenerated N-heterocyclic carbenes. *Electrochem. Commun.* **2012**, *25*, 116–118. [CrossRef]
56. Zhang, L.; Niu, D.; Zhang, K.; Zhang, G.; Luo, Y.; Lu, J. Electrochemical activation of CO_2 in ionic liquid (BMIMBF 4): Synthesis of organic carbonates under mild conditions. *Green Chem.* **2008**, *10*, 202–206. [CrossRef]
57. Yuan, X.; Lu, B.; Liu, J.; You, X.; Zhao, J.; Cai, Q. Electrochemical Conversion of Methanol and Carbon Dioxide to Dimethyl Carbonate at Graphite-Pt Electrode System. *J. Electrochem. Soc.* **2012**, *159*, E183–E186. [CrossRef]
58. Niu, D.; Zhang, J.; Zhang, K.; Xue, T.; Lu, J. Electrocatalytic Carboxylation of Benzyl Chloride at Silver Cathode in Ionic Liquid BMIMBF 4. *Chin. J. Chem.* **2009**, *27*, 1041–1044. [CrossRef]
59. Hiejima, Y.; Hayashi, M.; Uda, A.; Oya, S.; Kondo, H.; Senboku, H.; Takahashi, K. Electrochemical carboxylation of α-chloroethylbenzene in ionic liquids compressed with carbon dioxide. *Phys. Chem. Chem. Phys.* **2010**, *12*, 1953. [CrossRef]
60. Wang, X.Y.; Liu, S.Q.; Huang, K.L.; Feng, Q.J.; Ye, D.L.; Liu, B.; Liu, J.L.; Jin, G.H. Fixation of CO_2 by electrocatalytic reduction to synthesis of dimethyl carbonate in ionic liquid using effective silver-coated nanoporous copper composites. *Chin. Chem. Lett.* **2010**, *21*, 987–990. [CrossRef]
61. Feng, Q.; Huang, K.; Liu, S.; Wang, X. Electrocatalytic carboxylation of 2-amino-5-bromopyridine with CO_2 in ionic liquid 1-butyl-3-methyllimidazoliumtetrafluoborate to 6-aminonicotinic acid. *Electrochim. Acta* **2010**, *55*, 5741–5745. [CrossRef]
62. Feroci, M.; Orsini, M.; Rossi, L.; Sotgiu, G.; Inesi, A. Electrochemically Promoted C−N Bond Formation from Amines and CO_2 in Ionic Liquid BMIm−BF 4: Synthesis of Carbamates. *J. Org. Chem.* **2007**, *72*, 200–203. [CrossRef]
63. Lu, B.; Wang, X.; Li, Y.; Sun, J.; Zhao, J.; Cai, Q. Electrochemical conversion of CO_2 into dimethyl carbonate in a functionalized ionic liquid. *J. CO_2 Util.* **2013**, *3–4*, 98–101. [CrossRef]
64. Garcia-Herrero, I.; Alvarez-Guerra, M.; Irabien, A. CO_2 electro-valorization to dimethyl carbonate from methanol using potassium methoxide and the ionic liquid [bmim][Br] in a filter-press electrochemical cell. *J. Chem. Technol. Biotechnol.* **2015**, *90*, 1433–1438. [CrossRef]
65. Garcia-Herrero, I.; Alvarez-Guerra, M.; Irabien, A. Electrosynthesis of dimethyl carbonate from methanol and CO_2 using potassium methoxide and the ionic liquid [bmim][Br] in a filter-press cell: A study of the influence of cell configuration. *J. Chem. Technol. Biotechnol.* **2016**, *91*, 507–513. [CrossRef]
66. Feng, Q.; Huang, K.; Liu, S.; Yu, J.; Liu, F. Electrocatalytic carboxylation of aromatic ketones with carbon dioxide in ionic liquid 1-butyl-3-methylimidazoliumtetrafluoborate to α-hydroxy-carboxylic acid methyl ester. *Electrochim. Acta* **2011**, *56*, 5137–5141. [CrossRef]
67. Tateno, H.; Nakabayashi, K.; Kashiwagi, T.; Senboku, H.; Atobe, M. Electrochemical fixation of CO_2 to organohalides in room-temperature ionic liquids under supercritical CO2. *Electrochim. Acta* **2015**, *161*, 212–218. [CrossRef]
68. Abbott, A.P.; Dalrymple, I.; Endres, F.; Macfarlane, D.R. Why use Ionic Liquids for Electrodeposition? In *Electrodeposition from Ionic Liquids*; Wiley-VCH: Weinheim, Germany, 2008; pp. 1–13.
69. Smith, E.L.; Abbott, A.P.; Ryder, K.S. Deep Eutectic Solvents (DESs) and Their Applications. *Chem. Rev.* **2014**, *114*, 11060–11082. [CrossRef] [PubMed]

70. Abbott, A.P.; Frisch, G.; Ryder, K.S. Electroplating Using Ionic Liquids. *Annu. Rev. Mater. Res.* **2013**, *43*, 335–358. [CrossRef]
71. Prado, R.; Weber, C.C. Applications of Ionic Liquids. In *Application, Purification, and Recovery of Ionic Liquids*; Kuzmina, O., Hallett, J.P., Eds.; Elsevier BV: Amsterdam, The Netherlands, 2016; pp. 1–58. [CrossRef]
72. Peter, I. Effect of ionic liquid environment on the corrosion resistance of al-based alloy. *Key Eng. Mater.* **2017**, *750 KEM*, 97–102. [CrossRef]
73. Welton, T. Ionic liquids: A brief history. *Biophys. Rev.* **2018**, *10*, 691–706. [CrossRef]
74. Chen, L.; Sharifzadeh, M.; Mac Dowell, N.; Welton, T.; Shah, N.; Hallett, J.P. Inexpensive ionic liquids: [HSO 4] − -based solvent production at bulk scale. *Green Chem.* **2014**, *16*, 3098–3106. [CrossRef]
75. Endres, F.; Doughlas, R.M.; Abbott, A.P. *Electrodeposition from Ionic Liquids*; Endres, F., Abbott, A., MacFarlane, D., Eds.; Wiley-VCH: Weinheim, Germany, 2017; ISBN 9783527682706.
76. Kang, X.; Liu, C.; Zeng, S.; Zhao, Z.; Qian, J.; Zhao, Y. Prediction of Henry's law constant of CO_2 in ionic liquids based on SEP and Sσ-profile molecular descriptors. *J. Mol. Liq.* **2018**, *262*, 139–147. [CrossRef]
77. Song, Z.; Shi, H.; Zhang, X.; Zhou, T. Prediction of CO2 solubility in ionic liquids using machine learning methods. *Chem. Eng. Sci.* **2020**, *223*, 115752. [CrossRef]
78. Zhang, X.; Zhang, X.; Dong, H.; Zhao, Z.; Zhang, S.; Huang, Y. Carbon capture with ionic liquids: Overview and progress. *Energy Environ. Sci.* **2012**, *5*, 6668. [CrossRef]
79. Crosthwaite, J.M.; Muldoon, M.J.; Dixon, J.K.; Anderson, J.L.; Brennecke, J.F. Phase transition and decomposition temperatures, heat capacities and viscosities of pyridinium ionic liquids. *J. Chem. Thermodyn.* **2005**, *37*, 559–568. [CrossRef]
80. Yuan, D.; Yan, C.; Lu, B.; Wang, H.; Zhong, C.; Cai, Q. Electrochemical activation of carbon dioxide for synthesis of dimethyl carbonate in an ionic liquid. *Electrochim. Acta* **2009**, *54*, 2912–2915. [CrossRef]
81. Oh, Y.; Hu, X. Ionic liquids enhance the electrochemical CO_2 reduction catalyzed by MoO_2. *Chem. Commun.* **2015**, *51*, 13698–13701. [CrossRef] [PubMed]
82. Jouny, M.; Luc, W.; Jiao, F. General Techno-Economic Analysis of CO_2 Electrolysis Systems. *Ind. Eng. Chem. Res.* **2018**, *57*, 2165–2177. [CrossRef]
83. International Renewable Energy Agency. Available online: https://www.irena.org/publications/2020/Jun/Renewable-Power-Costs-in-2019 (accessed on 22 May 2021).

MDPI
St. Alban-Anlage 66
4052 Basel
Switzerland
Tel. +41 61 683 77 34
Fax +41 61 302 89 18
www.mdpi.com

Materials Editorial Office
E-mail: materials@mdpi.com
www.mdpi.com/journal/materials

www.ingramcontent.com/pod-product-compliance
Lightning Source LLC
LaVergne TN
LVHW070545100526
838202LV00012B/379